The Aqua Group Guide to Procurement, Tendering and Contract Administration

Second Edition

Edited by

Mark Hackett and Gary Statham

with contributions from

**Michael Bowsher, John Connaughton,
Michael Dallas, Paul Morrell, Alan Muse,
Erland Rendall, Simon Rawlinson, Nick Schumann,
Andrew Shaw and Peter Ullathorne**

WILEY Blackwell

Library of Congress Cataloging-in-Publication Data

The Aqua Group guide to procurement, tendering and contract administration. – Second edition /
edited by Mark Hackett, Gary Statham ; with contributions from Michael Bowsher [and 5 others].
    pages cm
    Includes bibliographical references and index.
    ISBN 978-1-118-34654-9 (paperback)
1. Buildings – Specifications. 2. Construction contracts – Great Britain. 3. Letting of contracts –
Great Britain. 4. Construction industry – Management. I. Hackett, Mark, 1962- editor. II. Statham,
Gary, editor.
    TH425.A678 2015
    692 – dc23
                                                    2015008101

A catalogue record for this book is available from the British Library.

Wiley also publishes its books in a variety of electronic formats. Some content that appears in print
may not be available in electronic books.

Cover design by Workhaus

Set in 10/12pt, SabonLTStd by SPi Global, Chennai, India.
Printed and bound by CPI Group (UK) Ltd, Croydon, CR0 4YY

C9781118346549_131221

Contents

Part I
Briefing the Project Team

Chapter 1
The Project Team

Introduction

Since the first editions of the Aqua Group's books, the process of constructing and running a built asset has become increasingly complicated. From inception to completion, through site acquisition, design, tender, contract and construction, each stage of the process is time-consuming and can be considerably expensive. The need to optimise the process is of paramount importance and the best base from which to achieve this is proper and efficient team work. It is therefore vital that all members of the project team are fully conversant, not only with their own role but also with the roles of others and with the inter-relationships at each stage of the project. All members of the project team can then play their part fully and effectively, contributing their particular expertise whenever required.

The make-up of any particular project team will depend upon the scope and complexity of the project, the procurement route and the contractual arrangements selected. There are already many different methods of managing a project and, no doubt, others will be developed in the future. This chapter is set in the context of traditional procurement and, although not exhaustive, provides an indication of the principles involved and the criteria by which other situations can be evaluated.

Parties to a building contract and their supporting teams

The parties to a building contract are the employer and the contractor. Those appointed by these two will complete the *project team* which can include:

The Aqua Group Guide to Procurement, Tendering and Contract Administration, Second Edition.
Edited by Mark Hackett and Gary Statham.
© 2016 John Wiley & Sons, Ltd. Published 2016 by John Wiley & Sons, Ltd.

The design team

- *employer
- *architect
- *quantity surveyor
- *principal designer
- project manager
- structural engineer
- building services engineers
- sub-contractors

In addition, the employer may appoint:

- *clerk of works

The construction team

- *contractor and/or principal contractor
- *site agent (or foreman, described in the contract as the person-in-charge)
- sub-contractors

It should be noted, however, that only those marked with an asterisk are mentioned in the Joint Contract Tribunal (JCT) Standard Building Contract With Quantities 2011 Edition (hereinafter referred to as the 'SBC'). This list is not exhaustive and to it could be added planners, landscape consultants, process engineers, programmers and the like. Furthermore, some roles may be combined and roles such as the project manager or principal designer may be fulfilled by individuals, firms or companies from varying technical backgrounds.

Rights, duties and responsibilities

The SBC is comprehensive on the subject of the rights, duties and responsibilities of the employer, the contractor and the other members of the project team mentioned in it. Not all the members of the project team are mentioned in the SBC and those not mentioned will usually be given responsibility by way of delegation from those who are mentioned. The delegation of any duties and/or responsibilities must be spelt out elsewhere in the contract documents and this will usually comprise part of the bills of quantities.

Whatever the size of the project team, all members should be familiar with the contract as a whole and, in particular, with those clauses directly concerning their own work, so that the project can be run smoothly and efficiently. It should be noted that the duties comprise (i) discretionary duties; (ii) mandatory duties; and (iii) statutory duties.

The employer

The employer is referred to throughout the contract and is expressly required to perform specific duties. The vast majority of these duties are codified and are carried out by the architect/contract administrator on behalf of the employer. However, the employer, as one would expect, retains the important duty of payment to the contractor for works which are completed in accordance with the contract. Such is the importance of the payment provisions in the SBC, and indeed in all construction contracts, that failure to adhere to the provisions may lead to statutory repercussions against the employer.

The architect/contract administrator

The architect/contract administrator is named in the contract and, as the designation contract administrator suggests, is not only responsible for carrying out the design of the works but also for the vast majority of the administrative duties under the contract on behalf of the employer. The architect/contract administrator is also the only channel of communication for any un-named consultants with delegated powers. Historically, the architect was recognised as being the person responsible for administering the contract. However, since the designation 'architect' is a protected title under section 20 of the Architects Act 1997, it can be used in business or practice by only those with the requisite education, training and experience. In this regard, if a person happened to carry out the duties of administering the contract and that person was not entitled to practice as an 'architect', then he could be held to be in breach of the Act. The title 'contract administrator' was added in order that other professionals could administer the contract without fear of breaching the Act.

The quantity surveyor

Quantity surveyors are named in the contract and their principal duties are in relation to the payment provisions, value of the works including the value of variations and, if so instructed, ascertainment of any loss and/or expense suffered by the contractor as the consequence of a specified matter.

The principal designer

The Construction (Design and Management) Regulations 2015 (hereinafter referred to as 'CDM 2005'), which came into effect on 6 April 2015, introduced the 'Principal Designer' to the project team (having been previously referred to as the planning CDM co-ordinator in the Construction (Design and

Management) Regulations 2007). When a project is notifiable, the principal designer is appointed by the employer, pursuant to regulation 5(1)a of the CDM Regulations. Standard contracts such as the SBC make provision for the appointment of a Principal designer. For instance, Article 5 of the SBC identifies that the Principal designer is the architect/contract administrator unless such other person is appointed.

The Principal designer is required to:

- plan, mange, monitor and co-ordinate matters relating to health and safety during the pre-construction phase to ensure the project is carried out without risks to health and safety;
- liaise with the principal contractor regarding the contents of the health and safety file, the information which the principal contractor needs for preparation of the construction phase plan, and any design development which may affect planning and management of the construction work;
- assist the client in the provision of the pre-construction information required by regulation 4(4) and to provide such of that information as is relevant to designers and contractors as is necessary;
- ensure that designers comply with their duties under Regulation 9 and ensure that all persons working in relation to the pre-construction phase cooperate with the client, the principal designer and each other; and
- prepare and update as necessary the project 'health and safety file' and, at the end of the construction phase, pass that file to the client; and

The clerk of works

The clerk of works may be appointed by the employer to act as an inspector of the works, solely under the direction of the architect/contract administrator. Traditionally, the role would have been taken by an experienced tradesman such as a carpenter, joiner or bricklayer. However, with today's highly complex and high-tech buildings, the architect/contract administrator, who will normally recommend the appointment, may need someone technically experienced or qualified and here the Institute of Clerks of Works will be able to assist in finding the right person. The clerk of works should be ready to take up the duties before the date of possession (how early will depend on the size and complexity of the project) and that person will either be resident on site or will visit the site on a regular basis during the period of the works.

The status of named consultants

While the architect/contract administrator and the quantity surveyor are expressly referred to in the contract and are expressly required to perform specific duties (many clauses include the phrase 'the architect/contract administrator shall'), they are not parties to the contract. Should the contractor have a grievance regarding the named consultants failing to carry out their duties prescribed in the contract, the only contractual recourse is to seek redress from the employer.

Unnamed consultants with delegated powers

The project manager, the structural or any other consulting engineers are not referred to in the contract and nor do they have any express powers under the contract. They do, however, have a duty, as the employer's persons, not to impede the progress of the contractor. Their position within the project team depends on the agreement they have with the employer or the architect. Where they have been given responsibility by way of delegation, perhaps for design or site inspection, they should be named in the contract documents and the extent of their delegated responsibility should be defined so that they have contractual recognition. Since they have no powers under the contract, if they need to issue instructions then this must be done through the architect/contract administrator.

As with the named consultants, if the contractor has any grievance against an unnamed consultant, the only contractual recourse is to seek redress from the employer.

The project manager

A project manager, who may be considered as an employer's representative, is likely to be appointed at the outset by the employer to whom he is directly accountable. The project manager is likely to be responsible for the programming, monitoring and management of the project in its broadest sense, from inception to completion, to seek a satisfactory outcome. This will involve giving advice to the employer on all matters relating to the project, and may include the appointment of the architect, quantity surveyor and other consultants. However, since the project manager is not mentioned in the contract, the project manager's position in respect of the contract works, which will be only a part of his overall duties and responsibilities, must be clearly determined and described in the contract documents. The duties and responsibilities delegated must not exceed those set out in the contract in respect of the employer.

The principal contractor

Whilst a main contractor may feature as a person in the design team for certain elements of the works, the contractor is usually appointed to construct the works. That same contractor is usually appointed as the principal contractor whilst carrying out the works. Under Article 6 of the SBC, the principal contractor is stated to be the contractor or such replacement as the employer shall appoint as the principal contractor pursuant to regulation 5(1)b of CDM 2015. The definition of a principal contractor in CDM 2015 allows for some flexibility but for the majority of construction contracts carried out under the SBC, the main contractor will be appointed as the principal contractor.

Since a principal contractor must always be appointed while work is in progress on-site, problems could arise where enabling works contracts, such as demolition or piling works, are required before the commencement of the main contract, or where a fitting-out contract follows completion of the main contract. In these circumstances, the employer will probably appoint a succession of principal contractors but care will need to be taken to ensure responsibility passes properly from one principal contractor to the next. In the event that the employer (client) fails to contractor, Regulation 5(4) provides that the employer must fulfil the duties of the principal contractor.

The principal contractor's duties are extensive and they are contained within three Regulations

12 – Construction phase plan; and health and safety file.
13 – Duties of a principal contractor in relation to health and safety at the construction phase.
14 – Principal contractors duties to consult and engage with workers.

These duties can be summarised as follows:

(Construction phase plan and health and safety file)

1. Draw up a construction phase plan which sets out the health and safety arrangements and site rules; ensure that the plan is appropriately reviewed, updated and revised from time to time.
2. Provide the principal designer with any information in its possession relevant to the health and safety file and, where the health and safety file is passed to the principal contractor (on conclusion of the principal designer's duties) the principal contractor must ensure that the file is appropriately reviewed, updated and revised to take account of the work and any changes that have occurred.

(Duties of a principal contractor in relation to health and safety at the construction phase)

3. Plan, manage and monitor the construction phase and coordinate matters relating to health and safety during the construction phase to ensure that, so far as is reasonably practicable, construction work is carried out without risks to health or safety.
4. Organise cooperation between contractors.
5. Coordinate implementation by the contractors of applicable legal requirements for healthand safety.
6. Ensure that employers and, if necessary for the protection of workers, self-employed persons apply the general principles of prevention in a consistent manner and follow the construction phase plan.
7. Ensure that a suitable site induction Is provided.
8. Ensure that the necessary steps are taken to prevent access by unauthorised persons to the construction site.
9. Ensure that welfare facilities that comply with the requirements of Schedule 2 of the Regulations are provided throughout the construction phase.

10. Liaise with the principal designer and share information relevant to the planning, management and monitoring of the pre-construction phase and the coordination of health and safety matters during the pre-construction phase.

(Principal contractor's duties to consult and engage with workers)

11. Make and maintain arrangements which will enable the principal contractor and workers engaged in construction work to cooperate effectively in developing, promoting and checking the effectiveness of measures to ensure the health, safety and welfare of the workers.

12. Consult workers or their representatives in good time on matters connected with the project which may affect their health, safety or welfare, in so far as they or their representatives have not been similarly consulted by their employer.

13. Ensure that workers or their representatives can (with certain limited exceptions) inspect and take copies of any information which relate to the health, safety or welfare of workers at the site.

Sub-contractors

Sub-contractors may feature as members of the design team as well as members of the construction team. This is because, in addition to carrying out work on site, they are often involved in the design and planning of specialist works in advance of the appointment of the main contractor. All sub-contractors employed by the contractor under the SBC have 'domestic' status and are fully accountable to the contractor. The contractor is fully liable to the employer for the sub-contract work completed and there is only a single exception in this regard. The exception is when the employer chooses a 'Named Specialist' to carry out certain of the works. If the employer opts for a Named Specialist then the employer will retain liability to the main contractor for any extension of time and loss and/or expense arising through the insolvency of a Named Specialist.

Statutory requirements

Under the SBC, the parties and the various consultants are charged with many duties, some of which are discretionary ('the architect/contract administrator *may ...*') and some of which are mandatory ('the contractor *shall ...*'). In addition, express provision is made for compliance with duties and responsibilities imposed by legislation as follows:

- Article 5 Appointment of principal designer and any necessary subsequent principal designer pursuant to regulation 5(1)a of the Regulations.
- Article 7 Refer a dispute to adjudication under Housing Grants, Construction and Regeneration Act 1996 (hereinafter referred to as 'HGCRA 1996').
- Clause 2.1 Comply with Construction Phase Plan – CDM Regulations and other Statutory Requirements.

- Clause 2.2 Contractor's Designed Portion – comply with Regulations 8 to 10 of the CDM Regulations.
- Clause 2.9 Construction information and Contractor's master programme – CDM Regulations.
- Clause 2.17 Divergences from Statutory Requirements.
- Clause 2.18 Emergency compliance with Statutory Requirements.
- Clause 2.19 Design liabilities and limitation – Defective Premises Act 1972.
- Clause 2.21 Fees or charges legally demandable under any of the Statutory Requirements.
- Clause 3.23 CDM Regulations
- Clause 4.6 Value Added Tax (VAT).
- Clause 4.7 Construction Industry Scheme (CIS).
- Clause 4.9 Interim payments – due dates and amounts due – HGCRA 1996 and Local Democracy, Economic Development and Construction Act 2009 (hereinafter referred to as 'LDEDC Act 2009').
- Clause 4.10 Interim Certificates and valuations – HGCR Act 1996 and LDEDC Act 2009.
- Clause 4.11 Contractor's Interim Applications and Payment Notices – HGCR Act 1996 and LDEDC Act 2009.
- Clauses 4.12–4.15 in respect of Interim payments – final date and amount, Pay Less Notices and general provisions and Final Certificate and final payment – HGCR Act 1996 and LDEDC Act 2009.
- Clause 6.1 Liability of Contractor – personal injury or death, employer's liability insurance – Employers' Liability (Compulsory Insurance) Act 1969.
- Clause 8.1 Meaning of insolvency – Insolvency Act 1986 or Bankruptcy (Scotland) Act 1985.
- Clause 8.6 Corruption – Bribery Act 2010 or Local Government Act 1972.
- Clause 8.7 Consequences of termination under clauses 8.4–8.6 HGCR Act 1996 and LDEDC Act 2009.
- Clause 9.2 Adjudication – HGCR Act 1996 and LDEDC Act 2009.
- Schedule 8 Health and Safety – Management of Health and Safety at Work Regulations 1999 and Health and Safety (Consultation with Employees) Regulations 1996.

These may change from time to time and may be augmented by the addition of Regulations issued by government under delegated legislation. Constant vigilance is therefore required to keep up to date.

The CDM regulations

The Construction (Design and Management) Regulations 2015 place duties on clients, principal desingers, designers and principal contractors to plan, co-ordinate and manage health and safety throughout all stages of a construction project. CDM 2015 came into force on 6 April 2015 evoking and replacing CDM 2007 which itself had consolidated a number of amendments to the

Construction (Design and Management) Regulations 1994 (hereinafter referred to as 'CDM94') made by the Management of Health and Safety at Work Regulations 1999, the Construction (Health, Safety and Welfare) Regulations 1996 and The Construction (Design and Management) (Amendment) Regulations 2000 (the latter amendments merely made minor amendment to the definition of designer).

It is worthy of note that contravention of the Regulations can be a criminal offence and serious offenders are likely to face prosecution. The Regulations apply to all construction work (as defined by the Regulations) but with only certain parts applying to some projects. The Regulations are in five parts and with supporting Schedules 1 to 5:

1. Introduction
2. Client duties
3. Health and safety duties and roles
4. General requirements for all construction sites
5. General

Clients are accountable for the impact of their approach on the health and safety in respect of those working on or affected by a project. Many clients, however, will know little about construction health and safety and as a consequence they are not expected to plan or manage projects themselves. Instead they may engage and rely (to an extent) on an expert such that they ensure that various obligations are completed, but they are not usually expected to do them themselves. If, however, the client fails to appoint a principal designer and/or a principal contractor, the client, under regulations 5(3) and 5(4), becomes responsible for fulfilling those duties. The client is also required to give notice in writing to the HSE if a project is "notifiable" under the terms of CDM2015. A project is notifiable if construction work is expected to longer than 30days and have more than 20 workers simultaneously at any point in the project or to exceed 500 persondays. In practical terms it is the principal designer (along with other designers) who has the most impact in the early stages of a project and, while such duties will continue throughout the design and construction periods, it is in the pre-contract stages that the majority of the designer's responsibilities remain to be discharged. The designer should seek design to reduce, if not avoid, risks to health and safety. Any design must consider the construction and subsequent occupational stages and the design documentation must include adequate information on health and safety. Such information should be included on drawings or contained in specifications and should be included in the health and safety file. The key duty of principal contractors is to properly plan, manage and co-ordinate work during the construction phase in order to ensure that the risks are properly controlled.

The HSE provides a free-to-download internet version of its guidance notes for the CDM2015 Regulations and a quick guide for clients, which provides practical advice on the application of the Regulations for all parties to whom they might apply.

Avoiding disputes

It is imperative that the project team implements formal procedures and proper documentation to succeed in the efficient and smooth running of building contracts – clarity and certainty are perhaps the key words. This is particularly relevant when seeking to reduce risk and avoid disputes. As projects become more complex, costs increase, margins tighten and employers demand greater financial and quality control, the margin between success and failure narrows and there is less flexibility to absorb the 'swings and roundabouts' which feature in the construction industry. The likelihood of disputes arising therefore intensifies.

The project team must be vigilant and ensure that the procedures employed and any working relationships built up produce an environment of co-operation rather than discord. Just as the project team strives for perfection in its working relationships, so the JCT strives to refine its contracts to take account of legislation, current practice and the latest decisions in the courts. Hence the frequent revisions to the JCT's standard building contracts which aim to achieve clarity and certainty in formalising the relationships between the parties.

In the event that a dispute does arise, the importance of proper documentation and compliance with formal procedures cannot be over-stated. If formal dispute proceedings become inevitable, it should be some comfort to know that proper documentation will be an asset rather than a liability.

On the assumption that the contract documents are complete, tenders are reasonable and reflect current prices, and that information is available when required and not subject to late changes on the part of the employer or the architect, the origin of a contractual dispute is seldom found to be in the dishonesty or incompetence of any party but rather in the failure of one member of the team to convey information clearly to another. Unfortunately, what is clear in the mind of the architect, for example, may be 'misty' to the quantity surveyor and 'foggy' to the contractor. This can lead to all sorts of problems! Conveying intentions and instructions clearly is vital for the successful management of a contract and, all too often, it is the breakdown or failure of communications that is the root cause of a dispute.

Communications

Myriad books, conferences and papers have been devoted to the subject of 'communications' and it is a matter that cannot be dealt with exhaustively here. However, set out below are certain golden rules to be observed by all members of the team in their dealings with each other:

- Do not unnecessarily tamper with the standard clauses of the SBC, but if the employer nevertheless requires it then employ a specialist.
- Ensure that the contract is executed prior to any start on site.
- Ensure that all team members have certified copies of the contract documents.
- Be realistic when completing the contract particulars and identify such information in the bills of quantities at tender stage.
- Issue all instructions to the contractor through the architect.

- Issue all instructions to sub-contractors or suppliers through the contractor.
- Use standard forms or formats for all routine matters such as instructions, site reports, minutes of meetings, valuations and certificates – preferably with sequential numbering.
- If instructions are issued other than in writing, ensure that they are confirmed in writing as soon as possible after the event.
- Ensure that everybody is kept informed, not just those who have to act.
- Be precise and unambiguous.
- Act promptly.

Examples of standard forms and suggested standard templates for the more important communications passing between members of the project team are given in later chapters.

Chapter 2
Assessing the Needs

The structure

The successful development of any building project requires an acceptance by all parties of an underlying structure or framework of operation. The RIBA Plan of Work 2013 published by RIBA Publishing and reproduced as Figure 2.1 is an excellent model for the briefing, designing, construction and occupational stages of a project.

As work proceeds, it is prudent to consider earlier decisions to ensure that the developing design continues to satisfy the employer's overall criteria. Generally, the procedures recommended in this book follow the RIBA Plan of Work 2013 under traditional procurement. However, the plan can also be used to suit procedures required by alternative non-traditional methods of building procurement.

The strategic definition

The strategic definition requires, as a core objective, one to establish the employer's business case, determine the strategic brief and consider other project objectives. The principal aim is to consider material issues prior to developing an initial project brief. At this stage of the process, one might consider issues regarding the assembly of the employer's project team and establish a preliminary programme regarding the duration of the project.

When considering the strategic brief, one is expected to assess the options for the employer with regard to issues such as its functional requirements, required quality, whole life costs and environmental requirements/constraints. When combined with the employer's business case, the strategic brief will enable one to consider if the project should comprise a refurbishment, an extension to an existing building or construction of a new building.

The Aqua Group Guide to Procurement, Tendering and Contract Administration, Second Edition.
Edited by Mark Hackett and Gary Statham.
© 2016 John Wiley & Sons, Ltd. Published 2016 by John Wiley & Sons, Ltd.

The RIBA Plan of Work 2013 organises the process of briefing, designing, constructing, maintaining, operating and using building projects into a number of key stages. The content of stages may vary or overlap to suit specific project requirements. The RIBA Plan of Work 2013 should be used solely as guidance for the preparation of detailed professional services contracts and building contracts.

www.ribaplanofwork.com

RIBA Plan of Work 2013

Stages / Tasks	0 Strategic Definition	1 Preparation and Brief	2 Concept Design	3 Developed Design	4 Technical Design	5 Construction	6 Handover and Close Out	7 In Use
Core Objectives	Identify client's Business Case and Strategic Brief and other core project requirements.	Develop Project Objectives, including Quality Objectives and Project Outcomes, Sustainability Aspirations, Project Budget, other parameters or constraints and develop Initial Project Brief. Undertake Feasibility Studies and review of Site Information.	Prepare Concept Design, including outline proposals for structural design, building services systems, outline specifications and preliminary Cost Information along with relevant Project Strategies in accordance with Design Programme. Agree alterations to brief and issue Final Project Brief.	Prepare Developed Design, including coordinated and updated proposals for structural design, building services systems, outline specifications, Cost Information and Project Strategies in accordance with Design Programme.	Prepare Technical Design in accordance with Design Responsibility Matrix and Project Strategies to include all architectural, structural and building services information, specialist subcontractor design and specifications, in accordance with Design Programme.	Offsite manufacturing and onsite Construction in accordance with Construction Programme and resolution of Design Queries from site as they arise.	Handover of building and conclusion of Building Contract.	Undertake In Use services in accordance with Schedule of Services.
Procurement *Variable task bar	Initial considerations for assembling the project team.	Prepare Project Roles Table and Contractual Tree and continue assembling the project team.	*The procurement strategy does not fundamentally alter the progression of the design or the level of detail prepared at a given stage. However, Information Exchanges will vary depending on the selected procurement route and Building Contract. A bespoke RIBA Plan of Work 2013 will set out the specific tendering and procurement activities that will occur at each stage in relation to the chosen procurement route.*			Administration of Building Contract, including regular site inspections and review of progress.	Conclude administration of Building Contract.	
Programme *Variable task bar	Establish Project Programme.	Review Project Programme.	Review Project Programme.	*The procurement route may dictate the Project Programme and may result in certain stages overlapping or being undertaken concurrently. A bespoke RIBA Plan of Work 2013 will clarify the stage overlaps. The Project Programme will set out the specific stage dates and detailed programme durations.*				
(Town) Planning *Variable task bar	Pre-application discussions.	Pre-application discussions.		Planning applications are typically made using the Stage 3 output. A bespoke RIBA Plan of Work 2013 will identify when the planning application is to be made.				
Suggested Key Support Tasks	Review Feedback from previous projects.	Prepare Handover Strategy and Risk Assessments. Agree Schedule of Services, Design Responsibility Matrix and Information Exchanges and prepare Project Execution Plan including Technology and Communication Strategies and consideration of Common Standards to be used.	Prepare Sustainability Strategy, Maintenance and Operational Strategy and review Handover Strategy and Risk Assessments. Undertake third party consultations as required and any Research and Development aspects. Review and update Project Execution Plan. Consider Construction Strategy, including offsite fabrication, and develop Health and Safety Strategy.	Review and update Sustainability, Maintenance and Operational and Handover Strategies and Risk Assessments. Undertake third party consultations as required and conclude Research and Development aspects. Review and update Project Execution Plan, including Change Control Procedures. Review and update Construction and Health and Safety Strategies.	Review and update Sustainability, Maintenance and Operational and Handover Strategies and Risk Assessments. Prepare and submit Building Regulations submission and any other third party submissions requiring consent. Review and update Project Execution Plan. Review Construction Strategy, including sequencing, and update Health and Safety Strategy.	Review and update Sustainability Strategy and implement Handover Strategy, including agreement of information required for commissioning, training, handover, asset management, future monitoring and maintenance and ongoing compilation of 'As-constructed' Information. Update Construction and Health and Safety Strategies.	Carry out activities listed in Handover Strategy including Feedback for use during the future life of the building or on future projects. Updating of Project Information as required.	Conclude activities listed in Handover Strategy including Post-occupancy Evaluation, review of Project Performance, Project Outcomes and Research and Development aspects. Updating of Project Information, as required, in response to ongoing client Feedback until the end of the building's life.
Sustainability Checkpoints	Sustainability Checkpoint — 0	Sustainability Checkpoint — 1	Sustainability Checkpoint — 2	Sustainability Checkpoint — 3	Sustainability Checkpoint — 4	Sustainability Checkpoint — 5	Sustainability Checkpoint — 6	Sustainability Checkpoint — 7
Information Exchanges (at stage completion)	Strategic Brief.	Initial Project Brief.	Concept Design including outline structural and building services design, associated Project Strategies, preliminary Cost Information and Final Project Brief.	Developed Design, including the coordinated architectural, structural and building services design and updated Cost Information.	Completed Technical Design of the project.	'As-constructed' Information.	Updated 'As-constructed' Information.	'As-constructed' Information updated in response to ongoing client Feedback and maintenance or operational developments.
UK Government Information Exchanges	Not required.	Required.	Required.	Required.	Not required.	Not required.	Required.	As required.

*Variable task bar – In creating a bespoke project or practice specific RIBA Plan of Work 2013 via www.ribaplanofwork.com a specific bar is selected from a number of options.

© RIBA

Figure 2.1 RIBA outline plan of work. Reproduced by permission of Royal Institute of British Architects.

The RIBA suggests that the initial project brief should comprise technical, managerial and design intentions and that research and development of the brief is likely to involve design team members in order to carry out:

- feasibility studies
- site or building survey and studies
- research into functional needs
- accessibility audits
- environmental impact considerations
- statutory constraints
- order of cost estimate(s).

Contribution to the initial project brief

After the initial euphoria in respect of the prospect of winning a job, members of the design team will appreciate that significantly more information will be required from the employer and from within their own organisation.

In order to consider any development of the initial project brief, each member of the design team will, typically, require the following:

- whether or not they are in competition with others for the commission
- the other members of the design team
- the status of the employer – contractor, developer, purchaser or owner-occupier
- the reason for the employer's desire to build – accommodation requirement, business expansion, company image or investment
- whether or not the employer has a site and whether or not any preliminary discussions have taken place with the planning or other statutory authorities
- any preliminary work undertaken by the employer to establish a need or a market for the proposed project
- any technical skills that the employer can contribute towards the project
- any critical dates/timing.

From within the design team's own organisations, each member of the design team will need to assess:

- whether or not they are experienced such that they can undertake the commission
- current workload – to determine whether or not the commission can be serviced by existing resources and/or new staff will have to be recruited
- the level of risk – there may well be an element of project development work for which no fee will be paid if the project does not proceed beyond a certain stage (for example, the design team will often be expected to undertake a degree of speculative work in assisting with the preparation of a site bid, particularly when the prospective employer is a developer)
- the effect of different procurement routes on their responsibilities and liabilities
- whether or not their professional indemnity insurance cover is adequate
- the employer's financial ability to complete the project.

The initial programme

One of the first questions the employer will ask will be: 'How quickly can I have my project completed?' At this very early stage it will be impossible for the design team to set a hard and fast programme. However, a simple bar chart (see Example 2.1) setting an early indication of the timescale for the various key activities will provide both the employer and the design team with a useful framework within which to try to operate. The chart needs to identify separately any activities where the time-scale is beyond the control of the design team and the employer. For example, this should comprise obtaining any development control and other statutory approvals.

At this stage, the employer has to rely on the professional judgment of his advisers in order to assess the likely implications or difficulties in respect of the project. The chart contained in this book has assumed a straightforward project which should not encounter any major delays.

In setting the initial programme, one should recognise that it is unlikely that the procurement route (be it drawings and bills of quantities, design and build or management contract) will have been established. Therefore, any assumptions which may have been made in the absence of firm decisions should be confirmed during the development of the brief and any adjustments should be made to the programme. Any assessment of the resources required for the project and an agreement to an initial programme will assist the design team when advising the employer in respect of the development brief.

The appointment

Traditionally, an employer's first appointment to the design team was the architect, who may then have assumed the role of project leader. However, there has been a significant rise in the use of professional project managers who are engaged specifically to control all of the main processes during a project. Whether an architect is appointed in the traditional manner or a project manager is the first consultant appointed, that person will be expected to advise on the appointment of the other consultants to the team, although those appointments are invariably made directly with the employer. In this way the architect/project manager avoids contractual liability for the professional performance of the other members of the design team and for the payment of their fees.

Most employers invite fee bids from consultants for the particular service required. However, even when the employer provides the fullest possible information, when few details of a scheme are certain the preparation of realistic fee bids at an early stage is both difficult and financially hazardous. Equally, the sensible evaluation of competing bids is difficult. Just as with the submission of an unrealistically low building tender, the employer should be cautious of the unrealistically low fee bid and question the consultant's ability to perform the required services satisfactorily – although in any competitive tender/fee bid situation, it is often the lowest price that is accepted. For those seeking to appoint consultants on the basis of quality rather than cost, the Construction Industry Council (CIC) published in 1998 *A Guide to Quality Based Selection of Consultants: A Key to Design Quality*. The publication states that the CIC 'firmly

Part I

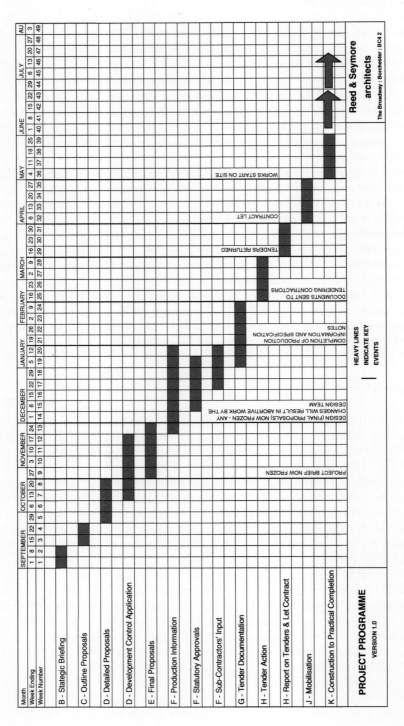

Example 2.1   The initial programme.

believes that it is neither in the best interest of the client [employer] nor of the project itself that consultants be selected on the basis of a procurement system which includes a price comparison of their professional services.'

Appointment documents

Certain professional institutions, such as the RICS and RIBA, make it compulsory for their members to provide written notification to their employer regarding the terms and conditions of their appointment. Any member of the design team who belongs to an institution that does not so require is advised to confirm, in writing to their client, at least the following:

- the name of the client
- the services that are to be provided
- the conditions of appointment
- how fees are to be calculated and when they are to be paid
- terms that will apply should the project change or not proceed.

Most professional institutions publish guides for clients relating to the appointment of their members, together with standard terms of appointment and forms of agreement for execution by the client and design team member. Typically these documents will also provide for the specification of the services to be provided, often by way of a standard tick list, and the fee to be paid for carrying out those services.

The British Property Federation (BPF) publishes a standard form of consultancy agreement for appointing professional consultants on construction projects. This standard form is one of very few forms to have been written from the perspective of the employer and, with some adjustments to each respective consultant's obligations, is designed to appoint all consultants, whatever their discipline.

Much of the criticism levied against standard forms of appointment is done so on the basis that each standard form is written to protect the liabilities of those seeking to impose it. Thus, the BPF form might be said to serve employers better than appointing an architect under a RIBA form.

The Construction Industry Council (CIC), which is a representative forum for professional bodies, research organisations and specialist business associations in the construction industry, publishes the CIC Consultant Conditions for appointing professional consultants.

The CIC refers to the conditions as being drafted with the aim of striking a fair balance between the interests of the client and the consultants. However, when the CIC Consultants' Contract is adopted for use on a project, it should be used for the appointment of all the consultants. This will mean that all team members are subject to consistent terms, will owe the same duty of care and will be under similar obligations in respect to co-operation, sharing of information and the co-ordination of design. Whilst, in theory, it is desirable to engage consultants on the same terms, using this form would mean that an employer would be faced with the task of trying to convince each different group that they should all carry a similar liability.

Collateral warranties

In addition to the normal appointment documents, consultants are frequently required to enter into collateral warranty agreements. Such agreements are used to create direct contractual relationships between parties other than the employer (e.g. funding institutions, purchasers, tenants) who are likely to suffer economic or consequential loss arising out of a construction project and any parties likely to cause such loss (e.g. consultants, contractors). The term collateral is used because the warranty is a side agreement to the warrantor's main contract with the employer.

The BPF and CIC both publish a series of standard Warranty Agreements. BPF standard Warranty Agreements comprise:

- CoWa/F (4th Edition 2005) – Collateral Warranty for funding institutions by a consultant.
- CoWa/P&T (3rd Edition 2005) – Collateral Warranty for purchasers and tenants by a consultant.

The CIC standard warranty agreements comprise:

- CIC/ConsWa/F – Collateral Warranty for funding institutions by a consultant.
- CIC/ConsWa/P&T – Collateral Warranty for purchasers and tenants by a consultant.
- CIC/ConsWa/D&BE (2nd Edition) – Collateral Warranty for employer by a consultant working for a design and build contractor.

The need for collateral warranties stemmed from the English law doctrine of privity of contract which provides that a person who is not a party to a contract cannot enforce that contract even if the obligations under the contract are for his/her benefit. On 11 May 2000 the Contracts (Rights of Third Parties) Act 1999 came into force and reformed the law relating to privity of contract by providing a person who is not a party to a contract with a right to enforce a contractual term. A third party may enforce a contractual term if the contract expressly provides that they may do so, or if a term purports to confer a benefit on the third party by reference to membership of a class or answering a particular description, even if not in existence when the contract is entered into. Whilst the Act is now a part of English law and, therefore, applies to any contract governed by English law insofar as the contract expressly states or purports to confer a benefit, it can be excluded from any contract by the use of a simple contracting-out clause. Many of the standard forms, whether they are consultant appointments or construction contracts, opt out of conferring third party rights. However, the JCT in the SBC affords users the opportunity to choose either third party rights or collateral warranties.

Whilst the Act, in theory, removes the need for collateral warranty agreements, this has not as yet happened in the construction industry – partly because of potential problems associated with the determination of the extent of a party's liabilities and the securing of appropriate insurance to cover them.

Chapter 3
Buildings as Assets

Simon Rawlinson

This chapter discusses how buildings can perform as assets for their owners and occupiers. Thinking of a building or a building portfolio as an asset encourages quantity surveyors and other consultants to concentrate on delivering outcomes for their clients.

An example of an outcome is an improved business performance resulting from an investment in a building project. A focus on outcomes contrasts with a traditional approach to consulting, which is concerned more with the delivery of services – such as the activities described in an appointment document, or a completed building. Business solutions that make better use of a client's built assets to improve performance are an important and adding-value aspect of consulting in the construction and facilities management industries.

Buildings as assets as well as buildings

Building consultants such as architects and quantity surveyors have an instinctive understanding of what a building is. They understand the importance of structure, thermal performance, comfort, durability and, above all, the appearance of the building and its effect on its neighbours, users and other stakeholders. Designers also understand how a building will be experienced by its occupiers, and in many cases how it will be used. These are high level professional skills. Being able to start from an accommodation schedule to deliver a high quality building that meets a client's brief requires a wide range of highly valuable skills.

Such is the technical complexity of the work which they undertake that building consultants have a tendency to think about the product of their work – reports, programmes, planning submissions and procurement strategy, all of which contribute to the successful delivery of the specified building. It is easy to overlook

The Aqua Group Guide to Procurement, Tendering and Contract Administration, Second Edition.
Edited by Mark Hackett and Gary Statham.
© 2016 John Wiley & Sons, Ltd. Published 2016 by John Wiley & Sons, Ltd.

the fact that most buildings have a business function and are in effect a solution to a particular business requirement – be it maximising retail sales in a particular catchment area, increasing passenger throughput in an airport or increasing the accessibility and availability of public services. The same can be said about some of the reports produced on a project – produced in support of a planning application or to obtain funding. 'Thinking about buildings' can sometimes result in the development of a narrow perspective concerning the most important aspects of a project. Taking the opportunity to think about the client's challenges, and the solutions that might be provided through an investment in buildings, is a great way of looking at a project from a different viewpoint, prompting new ideas whilst also ensuring that the value of the client's investment is optimised.

Single building or programme?

The first opportunity that clients might have to improve performance is the way in which they manage their building projects. Many of them will be occasional clients who will only ever build a single project. However, there are also a number of clients, including owners of large asset portfolios, who may have the opportunity to manage their capital investment and maintenance projects as a programme.

A programme of projects is typically a centrally coordinated portfolio of projects managed using common ways of working, performance standards and reporting processes. Projects in a programme do not have to be identical and nor do they have to be delivered by the same consultant and contractor.

The challenges that a client might face when delivering a programme of work include the complexity of managing multiple streams of work at different stages, the understanding of dependencies between projects and the effective prioritisation of investment. The client might also be missing out on opportunities to secure buying gain through joined up procurement, or to improve safety on projects by establishing and enforcing a common performance standard.

The advantages of programme-based management include greater visibility of budgets and programmes, simplified reporting and the ability to prioritise spending where it will deliver greatest benefit. There are many examples of clients and funding bodies that committed to more projects than they could afford, and equally, examples of clients who have not secured all of the value which they could from their expenditure. Programme management provides this opportunity. Examples of how programme management could provide benefits to clients with multiple projects include:

- Consistent management and reporting through a Programme Management Office.
- Joined-up procurement using the promise of repeat work to motivate contractor performance and volume of work to secure savings.
- Prioritised investment decisions based on the optimum mix of projects.
- Smoothed workload to make best use of contractor and client team capacity – known as 'feeding the machine'.
- Development of supply chain expertise through repeat work.

■ Development of a continuous improvement culture by all members of the project/programme team.

Programme management is often associated with large clients such as retail banks or utilities. However, it has also been used in the UK affordable housing sector to deliver home improvements associated with the Decent Homes standard, so it is an approach which can be adapted to different client requirements and scales. As clients in many industries join together in consortia to share resources and pool workload, the programme management approach will become increasingly relevant to clients in many sectors.

Buildings as solutions to business challenges?

Clients and more particularly their consultants can have a tendency to focus on the bricks and mortar aspects of a building project rather than the wider impacts it could have on the client organisation. Clients often find plans and elevations difficult to interpret and visualise, and it is equally challenging to understand the wider impacts that a well-designed and managed building project can have. These impacts can manifest themselves in many ways. The skill of the project team is to identify the opportunities and to help the client maximise the opportunity. Examples of these opportunities include:

■ Brand and culture. A building can say a great deal about the organisation which occupies it, but the messages need to be aligned. Civic buildings, such as town halls, are increasingly designed to support wider changes in service delivery models and should be welcoming and accessible to the wider public. In the current austere times, they also need to demonstrate that the client can manage money prudently and effectively. Similarly, clients who put sustainability at the heart of their business can use their building investments to demonstrate that green issues are central to their business thinking. In both instances, the contribution of the project team could be to ensure that design and performance of the building are wholly aligned with organisational brand and culture.

■ Supporting an organisation's business process. Designing buildings to facilitate the operation of the occupier sounds like an obvious solution but it is not something which is always undertaken as thoroughly or as radically as it could be. There are good reasons for this. Organisational standards might stifle innovation, or the project team may not have the time or the access to stakeholders to gain a full understanding of workflow issues and opportunities. Examples of how business process can be facilitated by solution-based thinking include optimised adjacencies in health buildings, improved sightlines in prisons that might reduce the number of warders needed or the design features used to encourage interaction in offices – including breakout space, cafes and even art.

■ Maximising the benefit of an investment. Private sector developers are clearly motivated by maximising the return on effort and capital employed. However, there are many examples of clients in the private and public sectors who may not be maximising the benefit that they could secure from built asset-related investment. Examples include clients who could secure greater disposal value for assets if they invested in targeted decontamination. In this instance, clients who invest in investigation and data capture in connection with a contamination problem may be able to use a low cost in-situ technique, which makes reuse of the land a viable option. In a very different benefit context, some clients secure improved performance from their consultant and contractor teams by increasing their use of contract incentivisation. Unfortunately clients' own processes can often get in the way of optimising investment benefits – resulting in delays in approvals, for example. Project teams can, and for that matter should, help clients to maximise the benefit that they secure from a project by understanding the client's business drivers, by supporting value and risk management and by bringing to the client's attention opportunities to use the project to improve business performance.

- Obtaining Capital Allowances. Certain items of plant and equipment can qualify for tax relief such that their selection may be attractive in terms of the subsequent tax breaks. This can lead to product selection being based more broadly than on capital cost alone, since the tax breaks achieved through the client's revenue stream must also be taken into account. The previous edition of this book contained a chapter on Capital Allowances but, for reasons of space, that chapter is not included in this edition though it is available on the book's companion website.

- Harnessing the energy of a project for the benefit of the client. Many construction investors are only ever occasional clients of the industry. Office tenants are a good example of a client whose core business is not construction. These clients clearly need support to guide them through unfamiliar construction processes. They also could benefit from advice on how to use a project to transform their business – whether by accelerating the skills development of members of the management team, or by using the project to reinforce other aspects of business change – such as the introduction of new ways of working in the completed office space. Project teams can help in this respect by thinking beyond the scope of the immediate commission – and by providing support at key moments such as handover and initial occupation. The 'soft landings' approach, where the project team undertakes post occupancy review and provides support to the client during the first 2–3 years of occupation, is a good example of this.

Everyday solutions-based thinking

Many of the examples of solutions-based thinking presented in this chapter have been strategic, looking at initial design concepts such as departmental adjacencies, procurement ideas such as incentivisation or optimised investment in decontamination to increase land value. Can a similar approach focused on outcomes be adopted on more routine activities?

The answer is almost certainly yes. The construction industry invests a huge amount of effort in working in ways which many clients find difficult to understand and, in some cases, difficult to see where value is being added. Creating documents with the end use in mind is a good place to practice output-based thinking. Simple examples that can be applied at all levels of industry include:

- Preparing reports that meet the client's needs. Ensure that the client's full requirement is understood and included, for example, inclusion/exclusion of Value Added Tax (VAT) in cost plans/cost reports.
- Ensuring that key messages are clear and well signposted. A client might have commissioned a surveyor to produce a 300 page NRM1-compliant cost plan, but the most important output may be the 2–3 page summaries which are circulated to stakeholders – such as funders or a Project Board. These summary documents need to be absolutely clear and aligned with the client's purpose.
- Manage change effectively. In addition to mitigating the effects of change, clear reporting is an important output for many clients. The production of reconciliation statements which summarise key changes alongside cost and performance impacts is another good example of using an outcome to improve existing processes.

Summary

Construction is but one process in the life of a built asset. If one is to satisfy a client's requirements, it is vital to remember this.

Thinking about how buildings perform as assets for their owners and occupiers provides vital insights into how project outcomes can be enhanced. In particular, thinking of a building as an asset encourages the design team to look at a project from different perspectives and to challenge solutions developed by the team.

Whilst there are many instances of strategic outcome-based thinking, the principles can be applied to everyday aspects of project delivery. Indeed, considering how any piece of work contributes to a project and how it will be used is a hugely valuable perspective that all consultants delivering built assets should develop.

Part II
Available Procurement Methods

Chapter 4
Principles of Procurement

Simple theory – complex practice

Procurement is the process of acquiring goods and/or services from a supplier or service provider. It is in essence a simple transaction but today's construction projects are often vast in scope, involve several designers, require specialist contractors, are executed at arm's length and take considerable time to complete.

A degree of complexity arises from the requirement to comply with numerous regulations, engage professional consultants, achieve value for money, demonstrate accountability, regulate complicated contractual relationships and achieve a timescale largely dictated by the client's specific business objectives. Nevertheless, irrespective of however complex modern construction projects may be, they are all based on the comparatively simple established principles of addressing cost, quality and time.

The eternal triangle

From a client's perspective, the COST, QUALITY and TIME paradigm might be considered as being the highest quality, at the lowest cost, in the shortest time. Unfortunately this is not always possible and a compromise has to be sought, based on the client's priorities. These criteria and the compromise can be visualised as a triangle (see Figure 4.1).

If these three factors are kept in balance, with appropriate quality being achieved at an acceptable price and in a reasonable timescale, the triangle would appear to be equilateral – with equal weight or emphasis being given to each factor. If, however, particular circumstances dictate that one of the factors must take precedence, at least one of the other two factors will 'suffer', or carry less

The Aqua Group Guide to Procurement, Tendering and Contract Administration, Second Edition.
Edited by Mark Hackett and Gary Statham.
© 2016 John Wiley & Sons, Ltd. Published 2016 by John Wiley & Sons, Ltd.

Figure 4.1   The procurement triangle.

Figure 4.2   Variations on the eternal triangle, showing the different priorities.

weight or emphasis. The decision as to the weight or emphasis to be given to each factor must lie with the client.

If quality is of paramount importance, adequate time must be allowed for the design and the specification to be perfected. In this regard, cost could rise on both counts. If speed of completion is paramount, then quality and cost may both have to suffer. If lowest cost is the priority, time may not be prejudiced but quality could suffer (see Figure 4.2).

The effect of these different priorities is relative and there is no reason as to why, with proper planning and management, those elements with a lower priority cannot be adequately controlled.

It is advisable to apprise the employer of the procurement options, recommend the most appropriate route, describe how the priorities are protected and outline the various implications. Only when the method of procurement has been determined can the pre-contract programme be finalised. This also represents the point in time when the specific services required of each member of the design team can be defined, fee proposals confirmed and agreements finalised.

One interpretation of the time, cost and quality priorities which may be derived from three different procurement routes is given in Example 4.1. It must be stressed that the particular circumstances of each project may result in different conclusions as to the procurement route that best satisfies the priorities of time, cost and quality. For example, a high level of provisional sums within a contract could undermine the cost certainty otherwise afforded by a particular procurement route. Additionally, it should be noted that, for ease of presentation, time allocation for building control approval has been omitted from Example 4.1. This element in itself can influence the choice of the procurement route.

A key decision when selecting a procurement strategy is based on the manner in which the detailed design is progressed. For example, when following a traditional procurement route the design team will develop the design, whereas with design and build procurement the design team may only prepare a design brief with the design itself being developed and completed by the contractor. Each procurement option has a different time, cost and quality scenario (the procurement triangle).

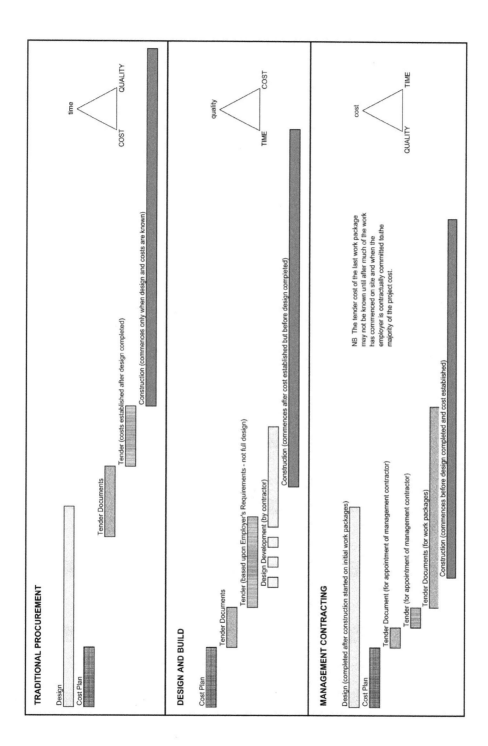

Example 4.1 Procurement options.

The importance and weight given to each criterion is established in the briefing process and is a fundamental prerequisite to deciding which procurement route to embark upon. Having established a client's priorities in respect of its criteria, even the best procurement strategies can be undermined by such things as:

- overly prescriptive briefs which give little flexibility to the design team;
- unnecessarily high standards;
- delays in decision making;
- the use of one-off solutions rather than standardised products;
- scope changes beyond those considered at the time of selecting the initial procurement route;
- poor communication in the supply chain;
- conflicting agendas and objectives;
- doubling up of professional resources (as between the client and the contractor);
- fixed mindsets and recycling old solutions; and/or
- issue of design information which is late and/or inadequate and/or incomplete.

It is of paramount importance that the foregoing risks are recognised and to have, as part of the procurement strategy, management systems aimed at their eradication or, at least, their control. The selection of a procurement strategy also influences the choice of which form of construction contract is used during the project. There are many standard contracts available to complement any chosen strategy. It is worth noting, however, that the selection of any particular form of contract is a secondary consideration to selecting a procurement route. The selected form of contract merely provides the administrative framework through which the procurement process can be achieved.

Other considerations

Cost, quality and time are not the only criteria to be considered when selecting a procurement strategy. The design team should make the client aware of all the other factors which may influence the decision as to the most suitable procurement system, such as:

- The Construction (Design and Management) Regulations 2015 (CDM 2015);
- risk; and
- accountability.

The Construction (Design and Management) Regulations 2015

Chapter 2 has already explained that CDM 2015 places statutory duties on clients, principal designers, designers and contractors to plan, co-ordinate and manage health and safety throughout all stages of a construction project. However, the principal contractor is responsible for health and safety during the construction phase of the project. The extent of the roles and responsibilities of the contractor and employer respectively under CDM 2015 depends on the nature of each project, the make-up of the employer's organisation and the employer's ability to undertake certain obligations. It is this indirect influence that may affect the procurement strategy, rather than CDM 2015 dictating the procurement strategy itself. Items that indirectly affect a procurement strategy comprise:

- management of the project;
- nature of the project and information available;
- contractor selection procedure and his supply chain management;
- activities with high risks to health and safety; and
- the employer's rules for site operations (e.g. the employer operating under a nuclear licenced site).

A key benefit of CDM 2015 is that whilst the regulations are mandatory, they are adaptable to suit any procurement strategy.

Risk

Risk is defined and explained in detail in Chapter 7. At this stage, it is sufficient to note that decisions on the allocation of risk between the employer and the contractor can have a fundamental effect on the choice of the right tendering procedure and contractual arrangement.

It is, therefore, important to establish the client's attitude to risk at the earliest opportunity. If the client is prepared to accept increased risk, the consultants will have much greater flexibility to devise an effective strategy based on non-traditional procedures.

Accountability

Accountability is closely linked with risk and is considered in Chapter 6. However, from the perspective of principles of procurement, decisions at an early stage may have a significant effect on tendering procedures and contractual arrangements. It is, therefore, essential that the client identifies the level of accountability required on the project as early as possible and for the consultants to explain the various principles involved, so that an early decision can be made.

Entering into the contract

Having decided which procurement route fits the client's brief, the process of entering into the construction contract may be broadly divided into three parts:

- deciding on the type of contract and the particular terms and conditions under which the work will be carried out;
- selecting the contractor; and
- establishing the contract price, or how the price will be arrived at.

Type of contract

The choice of the type of contract and the particular terms and conditions under which the work will be carried out will usually be made by the client in the light of advice received from his professional advisers and it will be inherently linked to the procurement strategy. The choice must be made at an early stage, as it will affect the way in which the contract documentation is prepared. In traditional single-stage competitive tendering, the type of contract and the actual conditions to be used must be defined and the method of establishing the contract price must be decided before tenders are invited. In negotiated contracts, these decisions can be delayed until the contractor is about to be appointed.

The range of contracts available for use is considerable and within each general type of contract a choice can be made as to the particular terms and conditions which are most suitable to the circumstances. There are two suites of contracts which cover most, if not all, domestic procurement strategies. These are the New Engineering Contract/Engineering and Construction Contract (NEC) family of contracts and the JCT 2011 suite of contracts. By way of example, and identifying the vast array of main contracts available, the JCT suite comprises:

- Standard Building Contract (SBC)
  - With Quantities (SBC/Q)
  - With Approximate Quantities (SBC/AQ)
  - Without Quantities (SBC/XQ)
- Intermediate Building Contract (IC)
  - With Contractor's Design (ICD)
- Minor Works Contract (MW)
  - With Contractor's Design (MWD)
- Design and Build Contract (DB)
- Major Project Construction Contract (MP)
- Constructing Excellence Contract (CE)
- Construction Management Appointment (CM/A) and Construction Management Trade Contract (CM/TC)
- Management Building Contract (MC), Management Works Contract Agreement (MCWC/A), Management Works Contract Conditions (MCWC/C
- Prime Cost Building Contract (PCC)
- Measured Term Contract (MTC)
- Framework Agreement (FA)
- Framework Agreement Non-Binding (FA/N)
- Repair and Maintenance Contract (RM)
- Building Contract for a Home Owner/Occupier (HOB)
- Building Contract and Consultancy Agreement for a Home Owner/Occupier (HOC)

To supplement the most widely used main contracts (namely, the SBC, DB, IC, MW and MP) the JCT also produces supplementary sub-contracts which provide further flexibility to the chosen procurement strategy.

Selection of the contractor – the tendering procedure

Selecting the contractor and establishing the contract sum, or how the sum will be derived, may be combined into a single operation or they may be two separate operations.

First it is helpful to look at a definition of tendering that has held good for a number of years:

> "The purpose of any tendering procedure is to select a suitable contractor, at a time appropriate to the circumstances, and to obtain from him at the proper time an acceptable tender or offer upon which a contract can be let."
>
> (The Aqua Group)

In fact, this is not so much a definition as a statement of the purpose of tendering. However, the sentence needs a degree of amplification in order to ascertain a full understanding of what it covers. The statement emphasises that the purpose of tendering is the selection of the contractor and the obtaining of a tender or an offer to carry out the works.

Although any tender must take account of the conditions of contract under which the work will be performed, the tendering procedure itself is not dependent on the type of contract to be used. The contract conditions have the same significance as, for example, contract drawings and bills of quantities, which together make up the total contractual arrangements. It is important to appreciate this; otherwise considerable confusion will prevail. Contractual arrangements are concerned with the type of contract to be entered into and the obligations, rights and liabilities of the parties to the contract. These may vary because of the type of project but there is no direct relationship between them and the tendering procedure.

For example, the same contractual arrangements could prevail in two instances where, in one instance, the contractor had been selected after submission of a competitive tender based on detailed drawings and bills of quantities, whilst in the other instance, he had been selected on the basis of a business relationship with the client and single tender negotiations. Chapter 24, Obtaining Tenders, contains a detailed explanation of tendering procedures.

Establishing price and time

Traditionally, there has been an underlying assumption that the tendering process was only concerned with establishing which contractor would submit the lowest price based on the design presented to him. However, greater consideration has been given to situations where the contractor is partly or wholly involved in the design; where circumstances require that construction commences before the design has been completed; or where, for any other reason, price and time alone may not be the only factors to be taken into account.

In considering the purpose of tendering, one must think in the broadest terms. A tender offer is not just a price but also a standard of quality and a time within which the work will be completed. The purpose of tendering may not be for one job; it may be necessary to consider it in terms of a total programme, of which that job is just one project.

The dynamics of tendering

Tendering procedures create dynamic situations. It is not correct to assume that there is a single 'procedure' which can be applied across the board on all construction works. In fact, 'procedure' is probably the wrong word but, as it is used so frequently, it may be misleading not to use it here. There are so many different factors to consider in each case and decisions to be made by the client and his advisors that it would be more appropriate to refer to it as 'the tendering structure and requirements'. For instance, tendering may comprise one of the following procedures:

- open tendering;
- single-stage selective tendering;
- two-stage selective tendering;
- selective tendering for design and build; or
- negotiation.

Much of what has been written with regard to past attitudes to tendering procedures also applies to contractual arrangements. There has been great emphasis on fixed price contracts with a somewhat poor realisation as to what the word 'fixed' entails. Frequently the assumption has been that the contract should be on a fixed price basis, unless the situation is such as to make this impossible.

Although the actual form and the conditions of contract are part of the contractual arrangements, this book is not concerned with the legal contract as such. The choice of the actual contract to be used on any given project is a matter to be decided in relation to the nature of the work, the needs of the client and the risks that the client is prepared to take in respect of the project.

Finally, although everything written here is in terms of a contract between the client and the main contractor, it is equally applicable to a contract between a main contractor and a sub-contractor.

Chapter 5
Basic Concepts

There have been two seminal reports in the last 20 years that have made recommendations to improve the performance of the construction industry in respect of the manner in which projects are procured and carried out. The two reports are (i) *Constructing the Team* by Sir Michael Latham; and (ii) *Rethinking Construction* by Sir John Egan.

Sir Michael Latham's report *Constructing the Team,* which was published in 1994, initiated a review of many of the traditional practices carried out in the construction industry. It highlighted the problems arising from industry processes that were uncompetitive, inefficient, expensive and adversarial. The report perceived that problems existed in respect of skill shortages, inadequate training and insufficient investment. It is easy to understand why the industry had such a poor image. The report also set an initial target of 30% cost savings against the cost of construction.

In Sir John Egan's report *Rethinking Construction* published in 1998, the cost reductions sought were raised to 40%. Whilst praising the industry for engineering ingenuity, design flair and flexibility, the report stressed the need for improvement through product development and by partnering throughout the supply chain. Partnering is considered in greater detail in Chapter 15.

Traditionally, only time and money were considered as being factors in selecting a contractor and then entering into a contract with him. Whilst this might be true in the broadest terms, there are many factors which affect either time or the financial outcome of the contract.

Although not every project will provide opportunities for the introduction of new techniques in terms of procurement procedure, the design team ought to review the possibility. Matters worth consideration include:

- the economic use of resources – labour, materials, plant and capital;
- the contractor's contribution to design and contract programme;

The Aqua Group Guide to Procurement, Tendering and Contract Administration, Second Edition.
Edited by Mark Hackett and Gary Statham.
© 2016 John Wiley & Sons, Ltd. Published 2016 by John Wiley & Sons, Ltd.

- production cost savings;
- continuity; and
- risk and accountability.

Economic use of resources

It may be thought that the money paid to the contractor is all that has to be considered when selecting a contractor. In this regard, traditional single-stage competitive tendering provides a good basis for monitoring this scenario. However, although the tendered sum provides a common basis for a comparison as to the extent of the resources to be used by the contractor, the client may be left with an unsatisfactory situation. If the tendering procedure and contractual arrangements are badly drafted, this may cause the contractor to use his resources inefficiently. To give a very simple example here: if a client has two similar projects, he may not necessarily obtain the lowest final price by setting up competition between one set of contractors for the first project and another set of contractors for the second and then taking the lowest in each case. If the two projects are linked or combined in one, the price of the larger single project may be less than if two projects were let separately. This might arise because of the savings in building resources and economies of scale.

The first priority in tendering is to ensure the most economic use of the building resources. One must bear in mind the particular needs of the client and then seek to ensure that the price paid for those resources is as low as is reasonably possible.

The economical use of contractors' resources is becoming a more difficult problem to assess compared to when the building industry was mainly craft-based and tendering reflected little more than the contractors' addition for profit and their management skill in organising the output of their craftsmen to complete the job. In such a situation there was little opportunity for contractors to vary the deployment of their resources. With limited plant and machinery available and with production knowledge limited to the best way in which to organise gangs of craftsmen to carry out the work, it was possible for the contractor to be given the quantities and for the lowest tender then to represent the lowest production resources.

In today's construction industry there is still much work that comes into this category but there are many elements in a modern building where different considerations apply. The situation regarding the supply chain is somewhat more complex as contractors have developed a preference to engage sub-contractors for a particular trade. The timing and co-ordination of sub-contractors' resources during the construction of a project remains a particular challenge to contractors.

Labour

If there is a general abundance of labour in the marketplace, a labour-intensive form of construction may be economical when managed correctly but if labour becomes scarce, difficulties may arise, leading to inflated prices. This general hypothesis applies to general labour and specific trades. It is also worth considering whether the emphasis can be shifted from being site-based, to factory-based labour carrying our prefabricated works – a cold, wet and windy site in the middle of January is not the best environment to encourage good productivity from the workforce.

Materials

When considering the selection of new materials, one should consider a number of factors such as their availability, bulk purchasing, relationships between contractors and suppliers and any special discounts available.

Life cycle costing studies have shown that increased investment in quality in original construction can reduce maintenance and replacement costs during the life of a building. This can lead to better economics overall and, therefore, increase the benefit of a project to a client.

Plant

When considering the use of a plant, a balance has to be struck between the use of larger factory-made components that may require an extensive plant such as cranage or smaller but more labour-intensive components. Should larger components be required, it is likely that a larger plant will be required to assist in placing the components (with the associated requirement for hardstandings for cranes and the like), but with a consequent saving in site labour. Continuity, or the lack of it, in the use of expensive items of plant has a significant effect on the economics of production.

Capital

Cash flow has a significant bearing on the success of most contracting organisations – keeping loan charges to a minimum can significantly affect contractors' tenders. Keeping retention percentages to a minimum and the prompt reimbursement of any agreed fluctuations in the cost of labour and materials can make a significant difference to the value of a tender. However, a balance has to be struck between the maintenance of a reasonable retention fund and the easing of a contractor's cash flow. Large international construction projects often provide contractors with advance payments in order to reduce the contractor's cost of capital. The use of advance payments is less prevalent in the UK construction industry.

There is, of course, another element that is of paramount importance in the quest for economical use of resources. The most careful planning of the use and co-ordination of labour, material and plant in the design stages of a project will be of no benefit unless the contractor has the management expertise to realise the master plan. This is particularly important in the planning and co-ordination of specialist sub-contractors' work. The aspect of project planning has led to the extensive use of pre-tender interviews with prospective contractors.

Part II

Contractor's contribution to design and contract programme

There are occasions when an experienced contractor can provide input to a design such that construction productivity increases and fewer resources are used. This helps to produce a more economical project. The importance of a contractor's contribution should not be understated. However, where sub-contractors are concerned, there is almost an undue readiness to consider the sub-contractor's contribution, whilst that of the main contractor is often accepted to a lesser degree. This dichotomy is partly due to the fact that sub-contractors tend to represent specialist activities entailing considerable design responsibility, whilst main contractors frequently show reluctance to accept any such responsibility. The main contractor's area of expertise is that of managing the construction process.

Engineering works differ somewhat from building works, in that the responsibility for detailed design is more frequently left in the hands of the contractor.

Cost planning techniques are usually employed to assist architects in designing economically. However, it is not always possible for the quantity surveyor or other members of the design team to know the best way in which the contractor's resources can be deployed to improve productivity in terms of cost and speed, particularly where a proprietary system is involved. For example, where a complete structure is to be fully designed before the contractor is chosen, a client may have to go out to the market and deal with specialist contractors in order to make use of a proprietary system. The installation of a proprietary system, such as a façade, is required to be integrated into the contractor's programme along with preceding and subsequent trades. Therefore, it is necessary to consider at the pre-tender stage whether a specialist contractor and/or a main contractor has a contribution to make and, if so, when his contribution should be introduced into the project.

Production cost savings

The construction industry's production methods are clearly a relevant factor in the cost of building and obtaining value for money for its clients. Each particular project is affected by a contractor's ability to put its production resources to economical use. Production resources are dynamic and production savings are continuously being made in the economy and the construction industry is no exception. It is the continued exploration of production cost savings that can provide one contractor a competitive advantage over another.

However, in the construction industry, methods of production can be limited by design and cost savings may be confined to more efficient methods of constructing a given design. It is difficult but not impossible to promote cost savings where a design has to be altered to achieve them. Legal liability for a design remains with the designer and this might partially explain why there is a certain reluctance to make any such amendments that could increase a designer's risk. However, there is no reason as to why a properly constituted value management exercise may not reveal some sort of cost saving. Ordinarily, unless contractual arrangements are devised, there will be no incentive for a contractor to look for productivity savings that affect the design.

In the event that there is potential for production cost savings during a project, it is essential that the client's advisors provide adequate time to incorporate a contractor's input and make arrangements to create an incentive for the contractor to attempt to make those cost savings. Although this may seem to be a daunting task, the client's advisors need only incorporate a form of bonus system into a contract and contractors are very familiar with the principles.

Continuity/framework arrangements often employ a form of bonus system in order to incentivise contractors. Continuity/framework arrangements are discussed in Chapter 14.

Continuity

Continuity of activity is perhaps one of the most important ways in which production and management resources can be used economically. The building industry has a unique problem in relation to the provision of continuity insofar as the majority of buildings are built on a one-off basis and on different sites. Nevertheless, a great many buildings are for clients who will want similar facilities on similar sites and, in this regard, some sort of framework/partnering approach would seem to offer the possibility of production cost savings.

Everyone takes longer to carry out a task the first time. Lessons can be learnt and there is a realistic possibility that savings can arise the second time and so on until, after carrying out the activity a number of times, maximum efficiency is achieved in production. A succession of one-off situations is bound to be challenging even when the most efficient management and production techniques are adopted, and unless very careful planning of production resources has been undertaken, the chances are that certain aspects of construction might prove uneconomical.

In order to make management and production savings, it would be necessary in this scenario to plan the procurement route and contractual arrangements for an initial project with a view to further projects being carried out by a single contractor. Obviously, productivity savings arising from such continuity must be translated into a corresponding benefit to the client. Framework agreements are dealt with in Chapter 14.

Risk and accountability

A client's ability to accept and manage risk is also an important factor to consider during the procurement procedure and when formulating contractual arrangements. It is sometimes thought that in shifting all of the risk onto the contractor, the issue of accountability has been satisfied. This is, however, a naïve view since the person best placed to handle a risk is the person who should be accountable for that risk.

The definition of risk in the Oxford Concise English Dictionary is '*a chance or possibility of danger, loss, injury, or other adverse consequences*'. It is important to realise that a contractor cannot be burdened with all risks in respect of a project. A client is burdened with a considerable risk by simply commissioning a building. A client is also burdened by the risk of matters such as what the authorities will permit him to build on the site. The design team take the risk in respect of the work they do. The question to be asked when deciding upon what procurement procedures and contractual arrangements should be employed is how should risk be apportioned between the contractor, the employer and the design team. It is not always appropriate for the client to ask the contractor to take all of the risk. It is likely to be more beneficial for an insurance company, whose very purpose is to take a risk, to charge a premium for certain risks. There is no doubt that in every contractor's tender there is a premium charged to the client for the risk he is burdened with. In certain instances, where the cost of a risk is very high but the probability of it occurring is low, it may not be worthwhile for the client to pass the risk to the contractor. Equally, it might be that in some instances, it will not be worthwhile to insure against a risk.

It is significant that the government does not pay premiums to insurance companies to take the risk of destruction of government buildings by fire. This seems to be good practice in terms of being seen to be accountable for public money. Such is the level of government resources, that there is no need to pay others to take on certain risks. If the government's position is contrasted with that of a small organisation, it is likely that the destruction of a substantial asset would be disastrous and it is better that someone else is paid a premium to accept the risk on that organisation's behalf.

When procuring a built asset, risk is very much affected by the extent to which a client's own resources can be deployed. It is often difficult to distinguish the risks that are appropriate for contractors to be burdened with because certain of the risks are inherent in their production and management activities, as opposed to a client's risks which are inherent in the project itself. It is, of course, entirely appropriate that contractors should take the risk for activities which are entirely under their control. In certain instances it will be appropriate for the contractor to be burdened with the risk of obligations which are also under the control of the client or his designers but in situations where the client might perceive that the contractor is best placed to manage a particular risk and that the contractor ought to be rewarded by including in his tender a premium for taking on the burden of that risk.

To illustrate this point, consider the conversion of an existing building where considerable risk is involved owing to the nature of the work. If a risk is perceived to be too great, it may be impossible to get the contractor to tender at all. If a risk is moderate then it may be open to a client with substantial financial resources to take the risk himself and let the contract on a cost reimbursement basis. On the other hand, an individual of limited means who wishes to carry out a conversion of his own house may feel it is essential to get the contractor to take the risk and, therefore, seek to obtain a fixed price even if to do so involves paying a premium on the price.

Fluctuation in the cost of labour and materials is a common risk to transfer to a contractor. However, prior to the global financial crisis, the developing overseas markets had caused a reduction in the amount of steel that was available to the UK construction industry. The cost of steel increased so sharply that in situations where contract had not been executed, contractors were unwilling to provide a fixed price. The reluctance of contractors to enter into a fixed price for the cost of steel was overcome by clients paying for steel on an actual cost basis plus a mark-up for overheads and profit.

Risk is one of the fundamental factors to consider when selecting suitable procurement procedures and contractual arrangements. It is a factor which depends on a client's attitude towards risk, its own financial circumstances, the nature of the risk, the perceived magnitude of loss if it were to occur and the probability of such a loss occurring.

Accountability

An early reference to accountability was made in the section above considering risk. Risk and accountability are two subjects that are very closely linked. However, accountability also has much wider implications and further discussion is set out in Chapter 6.

Summary

For convenience, set out below are the five fundamental factors to be considered in the selection of a contractor and the type of contract to be used. To some extent they overlap and they certainly inter-relate:

- the economical use of building resources;
- the assessment of the contractor's contribution in relation to the design and speed of construction;
- the incentives to make production cost savings and their control;
- continuity of work in all aspects; and
- risk and the assessment of who should be held accountable for it.

In addition, it is worth noting the distinction to be made between the selection of the contractor and the determination of the contract price. The traditional way of selecting a contractor by producing fully detailed drawings together with priced bills of quantities describing and quantifying the work to be done provides for the selection of a contractor complete with a breakdown of the contract sum. In this situation, the drawings and the bills of quantities are tender documents and they subsequently also rank as contract documents. Selection and determination of the price to be paid occur together.

In many other situations this may not be possible. What may sometime happen is that the contractor may be selected but on the basis that the contract sum is to be agreed at a later date. The final determination of the contract sum may take place many months later. Indeed, it does not have to be determined before the contract is entered into. If this point is appreciated, some of the more unusual methods of tendering will become easier to understand.

Finally, it should be emphasised that, although all these factors should be considered for every project or group of projects, it is recognised that in many of the smaller projects, some of the factors may not apply, particularly those projects which are completely one-off projects and which are not part of a larger general programme. Tendering procedures must be considered in the context of a national scene. The majority of building contractors are relatively small and the majority of builders would be unable and unwilling to tackle a job of any size. Many of the smaller contractors, whilst they may be capable of making a first class job when constructing to an architect's design, may neither wish have the ability to make the kind of contribution that either national or specialist firms can.

Chapter 6
Accountability

Accountability is a factor which can have a profound effect on the choice of procurement method.

Accountability arises whenever one party carries out an activity on behalf of another, such as a professional consultant (agent) carrying out work on the employer's behalf. The consultant must account to the employer for the actions that he takes. The extent of that account (or level of accountability) depends upon the employer's original brief and the degree of authority and responsibility delegated to the agent. In the procurement of a built asset, that authority and responsibility lie within a framework of established principles, with appropriate priority being given to cost, quality and time.

Background

Historically, there is great emphasis on accountability within the public sector, although it is just as relevant in the private sector. The difference between the two sectors relates to the reporting lines. In the private sector the relationship between agent and employer is direct: any matters can be settled between the two parties. In the public sector the responsibility of the agent to the employer (the public) is through an intermediary, usually an elected body, who will normally be advised by professional staff and who will ultimately be monitored by an independent audit body.

The accountability of the professional adviser is an important aspect of the tendering procedure in construction, because it is seldom possible for building owners to see what they are buying before it is built. Even when this may be possible to some extent, for example in the case of a standard building, each project is prototypical in nature due to the locality of its site.

The Aqua Group Guide to Procurement, Tendering and Contract Administration, Second Edition.
Edited by Mark Hackett and Gary Statham.
© 2016 John Wiley & Sons, Ltd. Published 2016 by John Wiley & Sons, Ltd.

Accountability differs from the other factors concerned with the tendering process, all of which are technical aspects, requiring professional advice to determine their effect. Accountability is a matter outside the strict construction process since it applies wherever an employer's agent is acting on the principal's behalf in respect of purchasing, for example, buildings, vehicles or stationery. Clearly, therefore, it is a matter on which employers should have an opinion and which should be considered to be part of their brief. Professional consultants, as agents, are responsible and accountable to the employer for both their actions and performance.

The modern concept of public accountability

The concept of public accountability is not a new concept. Earlier, production was craft-orientated and the sophisticated methods of modern production and construction were unknown. Problems related to production were insignificant. If the State voted to spend money on ships for the navy, the problem uppermost in mind was to ensure that the money voted for the ships was actually paid to the shipbuilder, rather than finding its way into somebody else's pocket. The method used in those days for the placing of orders was probably simple bargaining, implicit in which was the objective of value for money. The simplicity of the whole operation served to highlight only the points where the matter could go seriously wrong; these related entirely to graft and corruption.

As methods of construction have developed it has become more important to be able to show that value for money has been obtained. While the standard of honesty of public officials is generally high, the constraints imposed upon them to ensure that there is no corruption tend to work against the concept of value for money. Indeed, it is possible for all the strictest canons of public accountability to be adhered to and yet extremely poor value for money to be obtained due to the waste of resources inherent in present public accountability procedures.

The main reason for this is that, with complex technological production methods, the inefficient use of resources may easily arise if not fervently controlled. To avoid such waste, positive steps must be taken to ensure the optimum use of contemporary procurement and production techniques. These steps may be taken in value management/risk management workshops as covered in Chapter 7.

Contract documentation

The sums involved in purchasing construction work are usually considerable and this means that it is possible that money can be misappropriated – thus there is an increased opportunity for corruption than there is in many other areas of activity. Contract documentation for construction work is therefore very important and it is vital to the concept of obtaining value for money. The concept of being perceived as obtaining value for money may be complex and technical and it may vary according to the procurement route adopted (see also 'Dispensing with competition').

Proper price

Clients should have an assurance that they are getting value for money or paying the proper price for a project, even though it may not be possible to determine its total extent, or final price, at the time of entering into the contract.

When tenders are sought on a competitive basis with full contract documentation, the lowest price is normally accepted as being the proper price for the work since all tenders are related to a common base. Where full documentation is not provided (e.g. in plan and specification or design and build contracts), although the basis of tendering is still common, it can be more difficult to convince a client that the proper price is being paid, because the precise specification may not be clear at the tender stage.

Dispensing with competition

There are times when the use of competitive tendering is not possible, or will not necessarily produce value for money. There may be a number of reasons why the contractor should be selected without competition, and they may be complex and intangible. In many cases, it may be difficult to prove that there will be a monetary saving without the use of competitive tendering and the agent will have to convince the principal by justifying his professional judgment.

Where a contractor is selected on a basis other than equal competition, the contract sum or pricing mechanism must be negotiated and the matter becomes even more complex, especially if full documentation is not available. Not only does the selection of the contractor have to be justified, the concept of the proper price, or value for money, has also to be demonstrated.

Here again the documentation is vital – depending on the time and information available, a system must be set up that will not only allow a contract sum to be calculated but will also establish a pricing structure for the assessment of interim payments and variations where appropriate.

By their very nature, negotiated contracts often involve uncertainty, and the agent has a special responsibility to the principal to explain or justify the professional judgment that has led to the method of procurement recommended. In the public sector, whether in central government, local government or other government-controlled organisations, it can be difficult to make a convincing case when accounting for the decisions thus made.

Inflation

A further factor related to accountability arises from inflation. Particularly in times of high inflation the argument has sometimes been advanced that the tendering procedure should be arranged so as to bring forward the commencement of building in order that an earlier price level may be obtained (to beat inflation). This may be a real saving to a private client depending on his circumstances and the use to which his money is put in the meantime. To government, however, whose

Part II

concern is the percentage of resources devoted to building, there is no saving at all, as inflation affects both sides of the balance sheet. However, this fact needs to be weighed against other factors obtaining at the time; public finance accounting procedures and the way in which central and local governments organise annual financial allocations may well place a premium on early financial year starts within the public sector.

Value for money

Value for money, as a concept, may be defined as achieving the optimum use of resources – resources comprising money, manpower, time and materials, each of which may be regarded as equally important. Until comparatively recently money was often regarded as the most important resource in the public sector, the cheapest method of construction being the one most usually selected. With money becoming more 'expensive', the need to obtain more for a given amount became imperative, and this has tended to emphasise the equal importance of constituent resources. For example, if a contractor is overpaid £1000 on a contract through an accounting error, there is a monetary loss of £1000; if, through design errors which do not manifest themselves during the contract, there is a loss of £1000 of building labour and materials, then again there is a monetary loss of £1000. However, in the first case the money represents only financial resources which are no longer available for beneficial use, but in the second case the money represents not only a financial loss but also a loss of physical resources which have been used, wasted and lost.

Within both the public and the private sectors, the overpayment loss will be obvious; the loss of manpower and material resources may not. In either case, the professional advisers will be accountable to the building owner, to whom they are responsible, but whereas the former can be expected to be easily identified and recovered, the latter are considerably more difficult to detect and may never be recovered.

Government is concerned with the most economical use of resources by the country as a whole. More resources used in building each project mean fewer projects. While government is concerned with the monetary waste inherent in overpayment, this is very much the lesser of the two evils. Unfortunately, traditional systems only bring the overpayment to the fore, while the loss of resources remains obscure. Politically, too, overpayment is an easy matter to grasp and is highlighted in political debates as an example of government ineptitude, while the loss of resources is difficult to identify and for this reason is more often forgotten or ignored.

The definition of value for money has broadened over recent years. As money has become progressively dearer, so the concept has widened beyond that of obtaining best use of resources on site (usually by acceptance of the lowest tender) to embrace whole life cost of a project from initial procurement, through design and construction, to maintenance and cost in use and, perhaps, the cost of decommissioning in order that the best use/returns can accrue over the lifetime of the building. With this change in definition, terms such as 'life cycle costing' or 'whole life cost' and 'value engineering' have become the reasonable approach when purchasing a built asset, with value defined as 'the optimum balance of time, quality and cost'.

Summary

The problem, particularly with public building, is to identify whether or not there will be a saving in resources in relation to the way in which the contractor is selected. This is a complex and difficult task but in the long run the advantages to be obtained from many of the situations referred to later in this book will not be forthcoming unless the attempt is made.

Accountability is an extremely important matter in tendering procedures and contractual arrangements. It is essential that it is looked at in terms of all the resources used and not simply in terms of the price paid. The client's professional advisers must identify the effect that the procedure for the selection of the contractor may have on the use of all building resources and ignore, or certainly question, the historic traditions relating to public accountability.

Chapter 7
Value and Risk Management

Michael Dallas

Construction projects concern more than simply delivering buildings. They must (a) reflect the long-term business needs of those who commission and use them; and (b) deliver the benefits expected.

In this context 'value' lies in the effectiveness with which the benefits are delivered. Effective delivery requires that the supply chain clearly understands clients' long-term needs and delivers them efficiently and economically. Many initiatives focus on achieving a greater efficiency of delivery through the supply chain. 'Value management' complements efficient delivery by ensuring that efforts are deployed towards delivering the right buildings. It helps create affordable buildings that are productive and pleasant places to work in, that are easy to use and maintain and that contribute positively to the environment and communities in which they are situated. By contrast, effective 'risk management' ensures that value is not eroded by avoidable mishaps or uncertainties. To achieve this aim the team must, from the outset, ensure that conditions are put in place to enable successful project delivery.

This chapter describes how value management provides the means to articulate and deliver best value, whilst risk management provides assurance that the set-up and delivery of the project will avoid the erosion of value. It argues that both should be integrated with other project management processes to maximise the likelihood of a successful outcome.

Value management

Value management comprises a systematic process to define what value means for clients and end users of a facility, to communicate it clearly to the project delivery team and to help them maximise the delivery of benefits whilst minimising the use of resources. One of the most significant techniques used within the value

The Aqua Group Guide to Procurement, Tendering and Contract Administration, Second Edition.
Edited by Mark Hackett and Gary Statham.
© 2016 John Wiley & Sons, Ltd. Published 2016 by John Wiley & Sons, Ltd.

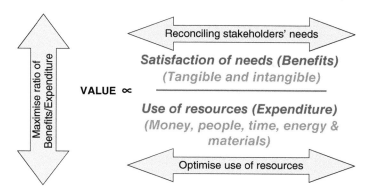

Figure 7.1 Optimising value for money.

management process is 'value engineering'. This is a structured technique that is applied to a design to deliver the required functionality at lowest cost and hence provide best value for money. Regrettably, all too often, value engineering is aimed purely at reducing costs, with little regard to the expected benefits.

At the outset of a project, value management provides an exceptionally powerful way of exploring clients' needs in depth, addressing inconsistencies and expressing these in a language that all parties, whether technically informed or new to the construction industry, can understand. This aspect of value management contributes significantly to developing the brief and building the project team (see Chapters 1 and 2).

Maximising value provides a mechanism for key stakeholders to reconcile their competing needs, optimise the use of resources (see the cost, time, quality triangle in Chapter 4) and then deliver most benefits for the expenditure of least resources. This balancing act is illustrated in Figure 7.1.

The process results in the following benefits:

- It defines what the owners and end users mean by value and provides the basis for making decisions throughout the project and is based on value. This is expressed through the use of a value tree or function diagram. The value tree establishes the relationship between and the hierarchy of so called 'value drivers' (those functions which are necessary and sufficient to deliver the project's objectives).
- Assessing the relative importance of the value drivers establishes the project 'value profile', providing the basis for a clear brief which reflects the client's priorities and expectations expressed in a language that all can understand. The resulting clarity in communications between all stakeholders provides a means for optimising the balance between differing stakeholders' needs so that each can understand and respect the constraints and requirements of the others.
- Assessing the scheme's performance against each value driver results in a 'value index', which allows the team to focus effort on those areas where there is most potential to add value.
- Estimating the resources needed to deliver each value driver (through function cost analysis) results in the 'value for money ratio' and provides a way of measuring value, taking into account non-monetary benefits and demonstrating that value for money has been achieved. This shows that the project is the

Table 7.1   Three phases in applying value management.

Project Stage	Focus of Activity
Inception to feasibility	Value articulation and project definition
Design and construction	Optimisation of benefits and costs
Commissioning and use	Learning lessons and performance optimisation

most cost-effective way of delivering the expected benefits to the business and provides a basis for refining the business case.

■ Value management supports good design through improved communications, mutual learning and enhanced team working, leading to better technical solutions with enhanced performance and quality where it matters. The methods encourage challenging the status quo and developing innovative design solutions.

Table 7.1 identifies three distinct phases in the application of value management to a construction project.

Value articulation and project definition

The use of functions to describe the benefits expected of a project enables the team to build direct links between the project objectives and the design solutions that later get built. These functions are referred to as 'value drivers' (see above) since they encapsulate what creates value for the client and end users. Essentially, building a function model of the project helps the members of the project team do three things:

■ They can agree on a clear description of what the project must achieve in terms of the benefits expected by the client and the end users.
■ They break this down into simple functional statements that describe the levels and quality of those benefits.
■ They break these down into clear statements that communicate to the design team those things that must be taken into account when the designs are developed.

Taken together, these statements describe the value that is expected at the outcome of the project and inform the development of the brief. They also provide the criteria upon which decisions should be based and feasibility stage options selected.

Optimisation of benefits and costs

Using the function model described above, the design team can develop designs that accurately reflect what the client and end users expect. However, satisfying the benefits in full represents only part of the optimisation of value.

To deliver good value for money the team must also make best use of the resources that are available. This is where techniques such as value engineering are useful. This technique builds upon the function model developed in the project definition stage to the point where it identifies the proposed building elements that are linked with each function. By adding the estimated costs of providing the elements that contribute to each function, the team builds up a picture of how much each function costs. By comparing this with the importance of each function, team members can assess where they are getting good value for money and where they are not.

The team generates ideas for performing the functions in different ways, starting with those that appear to offer lower value for money. Team members evaluate the relative merits of the ideas and develop those with most promise into detailed proposals for improvement. Finally, they submit their recommendations, often referred to as 'value improving proposals', based upon the proposals that they have developed, to the decision makers, who will decide which to include. The technique is quick and very effective, provided the team applies it rigorously and is not tempted to take short cuts.

Learning lessons and performance optimisation

Once the construction stage of the project is completed, the team should conduct a 'project review'. This provides the opportunity to check with the team and the building's users whether the full benefits that were defined at the outset of the project and predicted in the value engineering proposals have been realised. The review can explore what went well and what could have been improved and provide the opportunity for particularly successful generic value management proposals to·be 'banked' for use in future projects.

Table 7.2 identifies how the foregoing stages utilise the following value management study types. (See also Figure 7.2 for how these studies fit into the integrated process.)

The format for each study will usually follow the stages of a typical Value Engineering study, tailored to suit the project stage and the study objectives. Each study comprises four stages, identified in Table 7.3.

Table 7.2 The stages of value management.

No.	Project Stage	Typical Question Answered
VM 0	Inception	Is this the right project?
VM 1	Strategy	What are the project objectives and value drivers?
VM 2	Feasibility	What is the best option?
VM 3	Pre-construction	Is this the most cost-effective solution?
VM 4	Handover review	Did we achieve our expectations?
VM 5	Use	Is the business sustainable?

Part II

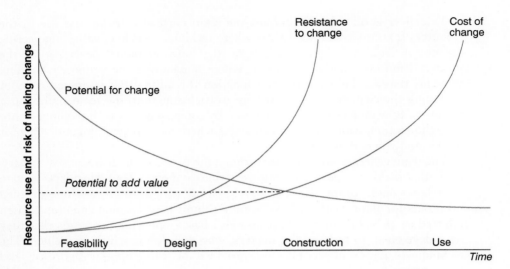

Figure 7.2   Opportunities reduce with time.

Table 7.3   Study structure.

Stage	Study Objectives
Preparation	Gathering and analysing information
Workshop	Workshop briefing, function analysis (and option selection, if appropriate)
	Creativity – generating ideas for improvement
	Evaluation – selection of the best ideas for development into proposals
	Development of proposals and action plans
Reporting and decision making	Presentation and reporting to obtain decisions on implementation
Implementation	Implementing the actions and proposals

Risk management

Risk management is a systematic process to identify, assess and manage risks in order to enhance the chances of a successful project outcome. Risk management can also identify opportunities to enhance value. Like value management, the focus of risk management studies evolves with the project. Table 7.4 illustrates this evolution.

Risk must be managed

Articulating and optimising value throughout the design development stages of a project is of little benefit if risks run out of control and manifest themselves in a manner which undermines the successful delivery of value. It is therefore necessary to set up a process by which risks may be managed effectively.

Table 7.4 Evolution of risk management studies.

Project Stage	Focus of the Risk Management Study
Inception	Are the risks acceptable?
Strategy	Are the conditions in place to proceed?
Feasibility	Are the risks allocated appropriately?
Pre-construction	Are the risks under control?
Use	What can we learn for the future?

In his report, *Constructing the Team*, proposing improvements to the construction industry, Sir Michael Latham stated 'No construction project is risk free. Risk can be managed, minimised, shared, transferred or accepted. It cannot be ignored'. Sir Michael Latham's report dates back to 1994 but his statement is no less fresh today than it was then. Indeed, it is necessary to take risk if one is to maximise the benefits (or value) in an organisation. The first and major benefit of risk management is that it enables senior management to embark upon projects with full knowledge that they will be able to control risk and thereby maximise rewards. Engaging in fire-fighting, whilst it may be exciting, is not an efficient way to control risk. It concentrates the attention on day-to-day matters whilst diverting attention away from the wider issues. Risk management helps the team to concentrate on the big issues and manage these in an orderly and timely way.

A formal risk management process delivers the following benefits for the project team:

■ It requires that the management infrastructure is in place to deliver successful outcomes. This includes setting clear, realistic and achievable project objectives from the outset.
■ It establishes the risk profile of the project, enabling appropriate allocation of risk, so that the party best placed to manage it has the responsibility for doing so. Risk allocation is a key component of contract documentation.
■ It allows the team to manage risk effectively and concentrate resources on the things that really matter, resulting in risk reduction as the project proceeds. It also enables the team to capitalise on opportunities revealed through use of the process.

Nothing ventured, nothing gained

The saying 'nothing ventured, nothing gained' neatly captures the principle that taking risks is necessary in order to gain rewards. Thus, project risk management is not about eliminating risk altogether but controlling the risks to which the organisation is exposed when undertaking a project. In order to control anything, one needs to understand it.

The first job in risk management is, therefore, to understand the project in depth. This is where the development of the value tree and derivatives (described above) neatly complement the risk management process. The following paragraphs explain the underlying concepts of risk management by going through the essential steps of the process.

Understanding the project

Listed below are some of the key questions that should be asked before embarking on any risk management study. In fact, these questions are the same as should be asked when embarking on a value management study. This is unsurprising since both must begin with a thorough understanding of the project.

- Why is the project being undertaken?
- What are its objectives?
- What are the benefits that are expected to flow from it?
- Who is involved and what are their interests?
- Is there an effective management structure in place?
- Within what business environment is the project being undertaken?
- What are the financial parameters relating to this project and are they realistic?
- What is the timescale and is it achievable?
- What are the constraints relating to the project?
- What are the factors that are critical to the project's success?
- What design solutions are being proposed and are they achievable within the timescale and budget?
- How much is it expected to cost?
- Is there enough money available to pay for it?

Risks

Having answered those questions, one should be on the way towards gaining a comprehensive understanding of the project. One can then begin to identify what could prevent benefits from being fully realised. These are the risks to the project.

Opportunities

At the same time as identifying what creates uncertainty in the delivery of any benefits, one also needs to understand the assumptions that have been made. These will usually have been embedded in the project business plan. Some of these could be overly cautious and, if the risk were not to materialise to the extent first anticipated, the outcome will actually be better than expected. These represent opportunities to increase project value and can be identified through risk management. Having identified such opportunities, it is wise to associate them with value management activities and thereby ensure that both techniques contribute towards increasing the project value.

Consequences

Having identified the risks, one needs to understand what would happen should those risks occur. These are the consequences of a risk occurring and are the things that could cause damage. The risk itself can remain a threat throughout the project but never cause any damage (apart from sleepless nights!).

Impact

It is only when the risk occurs that it will have an impact on the project. The impact may be financial, it may be related to the duration of an event or it may affect the quality of the product. It may, of course, affect all three. For each risk it is necessary to assess, qualitatively or quantitatively, the impact of its consequence on each of these dimensions.

Likelihood

In order to understand how any risk might be managed, one needs first to understand how likely it is that it will occur. Again, this can be expressed qualitatively or quantitatively – this could be qualitatively as a 'likelihood' in the absence of actuarial data or quantitatively as a 'probability' when actuarial data are available. Where something that may affect the project outcome is certain to happen, it is referred to as an 'Issue'. Issues need to be addressed by the project team in a similar way to risks.

Severity

The product of the impact and the likelihood of a risk provide a qualitative measure of the severity of the risk. This is generally sufficient to identify which risks require active management.

Risk management strategies

There are four major strategies for dealing with risk. These may be referred to in literature under various acronyms. Some use the acronym ERIC which is shown below in bold type and a commonly used alternative is TTTT which is also shown below (in italics).

- **Eliminate** (*Terminate*) show-stoppers and the biggest risks. This action will be taken only with the most severe risks which threaten to cause project cancellation or failure. For these risks it is essential that the risk be removed entirely or, at least, reduced to a bearable level such that one of the other strategies apply. There are several ways to do this. The team could review the project objectives and remove or re-think the objective which contains the intolerable risk: the result will change the project. The team could reappraise the whole concept of the project, again creating a changed project (with a different value and risk profile). It might be decided that the risk is so great that the project is not a viable proposition. This would result in the project's cancellation.
- **Reduce** (*Treat*) risks. This is achieved by undertaking surveys, redesign, use of other materials, use of different methods or by changes in the procurement plan. The approach is less drastic and more common than elimination. Reduction is accomplished by actions undertaken by members of the project delivery team.

For example, the design team can change the design to eliminate that bit of the design which causes the risk. Such a change might result in increased based cost. The team can undertake surveys to provide better information, thus removing uncertainty and thereby reducing the risk. Surveys will cost money to conduct. The team members can review different materials or equipment to achieve the required outcome. By adopting a more proven method or solution (which by definition is less innovative) the risk is likely to be reduced. They could propose different ways of working towards the project objectives and reduce risk in that way. They could package the work in different ways to reduce the interfaces between the trade contractors, thereby reducing the risk. The effect of reducing or treating risks is generally to increase the base cost and reduce the risk allowance because the team is paying to undertake the risk reduction process.

- **Insure** (*Transfer*) risks. Some risks can be insured or transferred to other parties. Insurance is a straightforward transaction where, in exchange for the insured party paying a premium, the insurer will meet the costs of any damage caused by the occurrence of an insured risk event. Another commonly used way of transfer is to allocate risk to the contractor through the contract. Allocating the risk in this manner does not actually reduce the risk. It normally results in the client paying a premium to transfer risk to the contracted party. Adopting this strategy may result in the client paying a premium to reduce his risk. The different types of contracts referred to in Chapters 8–15 allocate the risk between client and contractor in different but explicit ways. There is a common misconception that transferring a risk to another party in the project delivery supply chain eliminates the risk. This is, unfortunately, not the case. The contractor or other party to whom the risk is transferred must still manage the risk. The risk is still present. Should they fail to eliminate, reduce, insure or contain it, they must pay the consequences. If this is not done properly, the impact of the risk may still ultimately revert to the client.
- **Contain** (*Tolerate*) the risk within the unallocated contingency. This is the strategy to adopt for all minor risks. It involves no active management. The team judge that, should the risk occur, the cost is affordable within the general risk allowance or contingency fund allowed for the project. Because these risks are less severe, a competent management team will feel comfortable about addressing them in the event that they occur. Only then will they have an impact. However, should they occur, the full impact of the risk must be borne by the party to whom that category of risk is allocated through the contract.

Allocating management actions

Which strategy to adopt for the risks that have been identified will depend upon their severity (the product of likelihood and impact) and the team's 'risk appetite'. For each risk it is necessary to consider who is accountable should that risk occur. This person is normally called 'the account owner', and will be a senior manager or board member. The team must also decide who can best manage the risk, either on his own or in collaboration with others. This person is normally called 'the action owner'. Next, the team needs to consider what the action owner can do to implement any one of the strategies outlined above: this will be 'the management action'. Finally, the team needs to decide by when the action should be completed and when it should be reviewed.

Another consideration when allocating management actions is the proximity of the risk. If a risk is imminent, action must be taken urgently to mitigate it. If the risk is only likely to occur a long time into the future, mitigating action can be deferred to a more considered timescale.

In defining the action that the action owner should take, it is necessary to keep things in proportion in order to assess the resources needed to undertake the action and compare these with the impact should the risk occur. There is little point in expending more resource to manage a risk than the cost of its impact, should it occur.

If the project team is to maximise value and control risk, it is necessary to have effective value and risk management implementation plans in place. Essentially these should show individual responsibilities, how they communicate with one another, how proposals to improve value or reduce risk will be implemented, by whom and within what timescales. The plans will also set out the timetable for regular reviews and formal studies.

Value and risk management studies should be planned from the outset of any project. They should not be regarded as sticking plaster remedies to dig the team out of a hole when things have gone wrong. Sadly, this is often the way they are used, with the result that so-called value management exercises become little more than emergency cost cutting and in respect of which so called risk studies are merely reduced to fire-fighting plans. The earlier that risk and value management are started, the more benefits will accrue. Tables 7.3 and 7.4 illustrate the focus of value and risk management activities throughout the life of a project.

Figure 7.2 illustrates how the opportunities for adding value and reducing risk decrease with the passage of time.

Value and risk are complementary

Both risk and value management are needed to maximise the chances of project success. The reason for this lies in the different but complementary objectives of each discipline outlined above. Value is maximised using value management. Uncertainty, and consequent value destruction, is minimised using risk management. One might hold a project launch event in order to instigate the use of value and risk management. During an event, the project team would be invited to assess the extent to which the project is set up to achieve certain criteria.

Examples questions that will interrogate the likelihood of a successful project outcome are addressed in 'Understanding the project'.

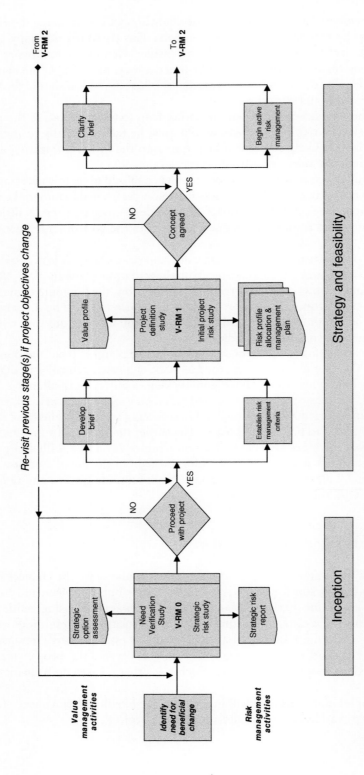

Figure 7.3 The integrated process of risk and value management.

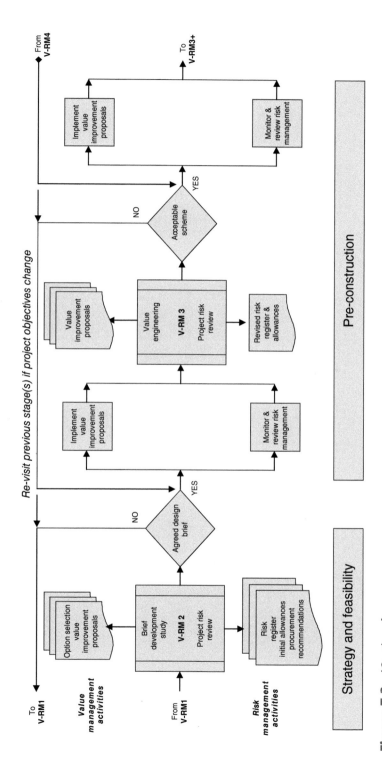

Figure 7.3 (*Continued*)

Part II

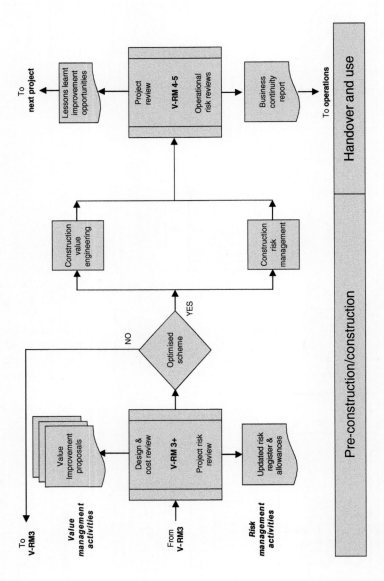

Figure 7.3 (Continued)

Similarities in the processes

Whilst the value and risk processes may differ in detail, they have the following similarities:

- the preparation stage, to understand the project and the issues relating to it;
- the requirement for consultation with and involvement of the main stakeholders;
- the use of facilitated workshops involving a balance of stakeholders, disciplines and characters;
- the development of value and risk profiles by which progress towards improving value and reducing risk may be assessed;
- the development of proposals to improve the project;
- the agreement of the necessary management actions to implement those proposals;
- the need for an explicit implementation plan;
- the written record, or report on the outcome, providing a clear audit trail; and
- the need for regular reviews to monitor implementation and report progress.

The integrated process

Combining the value and risk processes within a single programme of studies is therefore both logical and practical. Essentially, the integrated process comprises a number of formal studies that coincide with key milestones, or decision gateways, throughout the life of the project.

Between the formal studies, the progress of implementation and management actions should be reviewed by a responsible person on a regular basis and reported in the regular project reports. These reviews and progress reports are likely to be conducted and reported separately. This is because different people within the project team may be responsible for conducting them.

This process is illustrated in Figure 7.3 which shows the integration of the risk and value processes.

Further information on the application of value and risk management may be found in:

- UK Cabinet Office Best Management Practice portfolio, of which Management of Value (MoV®) and Management of Risk (M_o_R®) are parts and for which the APM Group, www.apmgroup.co.uk, are the official accreditors. (MoV®) and (M_o_R®) are published by TSO (the Stationery Office), 2010, ISBN 9780113312764 and 9780113312740, respectively.
- Value and risk management, a guide to best practice, Michael F. Dallas, Blackwell Publishing, 2006.

Part II

Chapter 8
Fixed Price and Cost Reimbursement

Fixed price and cost reimbursement are terms that describe the two methods of making payment. However, they are somewhat misleading because, although they indicate the principal methods employed for paying for work that is complete, very rarely is any contract discharged entirely by only one of the two methods. Usually a combination of both methods is employed and the description of the contract is determined according to that method which is predominant.

The principles, however, remain valid. Fixed price items may be defined as items paid for on the basis of a predetermined estimate of the cost of the work, including allowances for the risks involved and the effect of market forces in respect of a contractor's workload. The tendered price is paid by the employer, irrespective of the cost incurred by the builder.

Cost reimbursement items may be defined as items paid for on the basis of the actual cost of the work.

The JCT Prime Cost Building Contract 2011 (PCC) for building work provides a detailed definition of Prime Cost to the contractor under the heading 'Definition of Prime Cost' in schedule one. The definition of Prime Cost is further discussed in Chapter 10.

Fixed price

The term fixed price can apply to a unit rate, a section of work or equally to a whole contract. Similarly, it must be appreciated that a contract may consist of a multiplicity of unit rates, a series of elemental or trade sections or a single lump sum. Understanding this should dispose of the common misconception that a fixed price contract is a single lump sum contract.

A fixed price contract need not have a finite sum attached to it at the beginning of the contract. A schedule of fixed rates (without quantities) is a fixed price

The Aqua Group Guide to Procurement, Tendering and Contract Administration, Second Edition.
Edited by Mark Hackett and Gary Statham.
© 2016 John Wiley & Sons, Ltd. Published 2016 by John Wiley & Sons, Ltd.

contract, because the basis of payment has been predetermined – the price is fixed. Only the quantity of work is unknown and this is ascertained by measuring the work as it is done. This is still true even if the rate varies with the quantity of work completed. Fixed price is discussed further in Chapter 9.

Cost reimbursement

In this system, payment is not based on a predetermined contractual estimate of cost. A contractor is paid whatever the work actually costs him within the limits of the contractual arrangement. The contract will prescribe the rules or formulae in order to calculate the cost to the employer.

The essential difference between a fixed price and cost reimbursement contract is that in the case of a fixed price contract, a contractor will undertake to do the work at a price that he has estimated in advance. If he is incorrect in his estimation then he will suffer the risk of being wrong. In a cost reimbursement contract, however, the employer pays the actual cost to a contractor. This is often described as an 'open-book' approach. If the actual cost of the project is higher than any pre-contract estimate, then the employer automatically pays the additional cost. Equally, if the actual project costs less than the pre-contract estimate, then the employer gains from the savings.

Application to contract elements

Having established the principles, one can now consider practical elements that make up a contract and how the principles of fixed price or cost reimbursement apply in each case.

Unless the contract is for a single lump sum figure to supply a specified building, a 'fixed price contract' usually comprises the following elements:

1. preliminaries – often a series of individual sums such as site management costs, plant costs, site establishment costs, etc.;
2. unit rates – work fixed in place;
3. PC sums;
4. defined and/or undefined provisional sums – work anticipated to be required but not yet designed, including contingencies;
5. provisional sums – for work executed by a local authority or by a statutory undertaker executing work solely in pursuance of its statutory obligations; and
6. overheads and profit – sometimes included in preliminaries and unit rates but sometimes expressed separately as a sum or percentage in the summary.

One can identify that the first two items are of a fixed price nature whilst the third, fourth and fifth items embody the cost reimbursement principle. The sixth item, normally overheads and profit, merits further comment. Whether the contractor prices these from first principles or fixes his reimbursement as a percentage, the resultant sum becomes fixed upon the contract being entered into and the contractor will be held to this price commitment whether his actual costs are higher or lower. This does not, however, rule out the possibility of overheads and profit being reviewed in the event of variations or delays for which the employer is responsible. The point here is that the contractor's obligation is fixed in respect of those things which were known about at the time of entering into the contract.

Even in a cost reimbursement or 'cost plus' contract the allowance included for overheads and profit is committed in advance as a lump sum or as a percentage of the cost spent on the labour and materials. This is regardless of whether the actual cost of the overheads and management is higher or lower than this. There are, therefore, cost reimbursement items for labour, materials and plant but which also include a fixed price item for management. It may appear confusing to say that a percentage can be a fixed price in this context. It is said on the basis that whilst the underlying costs to which the percentage relates will fluctuate and the payment due in respect of overheads and profit will fluctuate in proportion, the percentage being applied is in itself fixed.

In practice, as previously explained, so-called reimbursement contracts frequently contain elements of fixed price and many fixed price contracts contain elements of cost reimbursement.

Fluctuations

An element of cost reimbursement that may come into otherwise fixed price contracts concerns the payment for fluctuations in the cost of labour and materials. A distinction is frequently made here by calling a contract where fluctuations are not paid a 'firm price contract'. When fluctuation clauses are included in a contract, an adjustment is made for the increase or decrease in the basic cost of labour and materials. Any such adjustment is dealt with on a cost reimbursement basis and is based on fluctuations in the market price. However, when a fluctuation in the cost of labour and materials is calculated on a formula related to national indices, contractors will be reimbursed at a level which may or may not reflect the actual price fluctuations that they encounter.

Target cost contracts

Having defined fixed price and cost reimbursement as being the principal methods of payment, that is not the end of the matter. It is possible to combine them both, not only by having part of the work paid for on one basis and part on the other but by applying both principles to the same items of work. In fixed price contracts, contractors take the risk of their estimates being wrong, whilst in the case with the cost reimbursement contract the employer is at risk in relation to the final cost. If, however, it is desired that the risk of the estimate should be split between the two parties then contracts can be devised in such a way as to allow this. A target cost contract may be employed. Target cost contracts are explained in detail in Chapter 11.

Use

The appropriate allocation of risk between the employer and the contractor will determine the extent to which fixed price and cost reimbursement will be employed in a contract. In assessing the situation there are two main issues to consider. Both issues relate to the circumstances of the employer and of the contractor.

The employer's position

The following criteria should be considered:

- the employer's financial position;
- the employer's time requirement for building work;
- the employer's corporate restrictions; and
- the work available in the market place.

In a situation where an employer has a defined budget for a project and adequate time for its professional advisers to obtain fixed price quotations, that route should be pursued in order to minimise the employer's risk. However, if time is more important than the capital cost of a project, the employer's position may be best protected by taking a greater financial risk and moving towards a cost reimbursement type of contract. It is, however, very important that an employer is advised on the correct balance between the two methods of payment. Any attempt to reduce an employer's financial risk without providing adequate information to a contractor will undoubtedly backfire and an employer may have to pay more as a contractor will price to cover potential risk which may not eventually arise.

Employers may have a predetermined method of procurement for building work which will override their best interests. For example, a local authority may require fixed price tenders to be obtained from an accountability point of view, yet this method may preclude the authority from having the work carried out during a particular financial year and it would be possible for it to lose the money allocated for the scheme.

When work in the construction industry is scarce, contractors will inevitably be prepared to accept greater financial risk in order to obtain work and the employer's professional consultants should advise on that basis.

The contractor's position

The following criteria should be considered:

- the contractor's financial position;
- the contractor's financially acceptable risk level;
- the extent to which the work is defined; and
- the work available in the market place.

As with an employer who has limited financial resources, contractors in a similar situation cannot afford to take large financial risks. In this regard, when a contractor is asked to take a risk, he will add a sufficient amount to his fixed price to cover the eventuality of that risk occurring. In certain situations, a contractor may not be prepared to take the risk at all and he may refuse to tender for the work. On the other hand, contractors with substantial financial resources are in a position to take on more financial risks and submit fixed price tenders without prejudicing their immediate financial stability. Obviously, a contractor must be satisfied that the risks he is prepared to take when pricing a project will even themselves out over the whole range of work that is to be undertaken.

The level of financial risk that a contractor is prepared to take is a commercial decision. Two contractors with similar financial resources may have different attitudes towards the level of risk that they can each assume. For example, one contractor might be prepared to estimate and shoulder the burden of any material and labour cost increases, whereas the other contractor might require reimbursement for any such increases. This action may preclude the latter from being invited to tender.

In instances where it is either difficult or impossible to define the work clearly, such as schemes incorporating some form of alteration or refurbishment, it could be beneficial for at least some of the work to be let on a cost reimbursement basis. Attempting to estimate a fixed price for work that is not properly defined will result in the items being heavily priced to cover certain elements of the risk.

Where new building techniques are employed, it may be in the interests of both parties to have the work carried out on a cost reimbursement basis, as the true risk may not be definable. Attempting to allocate a fixed price may, in these circumstances, result in too low or too high a price – with disastrous effects all round.

Market conditions can affect a contractor's attitude to tendering. When the building market is depressed, the contractor may take the view that he is prepared to take greater risks and tender on fixed prices. In a more buoyant market, he would be more likely to look for a cost reimbursement basis.

Programme

Having regard to what has previously been said in this chapter, one might appreciate that cost reimbursement contracts are most commonly used where completion in the shortest time is essential, or where it is impractical to define the work fully before it is executed.

Figure 8.1 shows a diagrammatic representation of the different sequence of events between fixed price and cost reimbursement contracts. The vertical divisions of the matrix are not intended to indicate any particular time interval – they are provided purely to assist visual comparison of the alternatives. The length of the activity bars are not necessarily significant, except in allowing a comparison to be drawn between the two systems. In practice, the duration of any activity must be agreed between the participants in order to achieve the employer's expectations.

Fixed price (single-state competitive tendering)

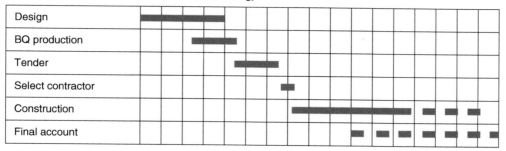

Cost reimbursement (negotiated tender)

Figure 8.1 Sequence of events – fixed price and cost reimbursement.

However, Figure 8.1 indicates that considerable time saving can be achieved using a cost reimbursement contract – not only in achieving completion of the work, but also in completing the final account. Since the cost verification process has to be commenced from the very beginning of the contract, if progress payments and cost reporting are going to be realistic, the final account should almost be complete when payment is calculated for the last period of work.

Summary

When there is sufficient time it is beneficial to carry out investigative work and produce as detailed a design and specification as possible. This work will reduce the risk element in the contract and influence the type of contract eventually used. On all contracts, payment is made either by fixed price or cost reimbursement: FP + CR = 100%. Almost all building contracts are a combination of both.

With an increase in the use of cost reimbursement and management type contracts which are often let on an 'open-book' basis, it is clear that employers are prepared to take more financial risk in pursuit of shorter procurement times and lower prices. Partnering arrangements are often set up on an 'open-book' basis – with a view to all parties achieving best value and sharing the financial risks and benefits on an equal basis, unlike the case with traditional fixed price contracts.

Part II

Chapter 9
Fixed Price Contracts

Fixed price or lump sum contracts provide for a predetermined estimate of the work made by the contractor prior to entering into the contract, irrespective of the actual cost incurred by the contractor. A fixed price contract might comprise:

- a single lump sum;
- a series of element or trade totals; and
- a multiplicity of unit rates.

Fixed price contracts account for the majority of building contracts and they are still considered as being the norm in terms of the number of contracts let, if not by value. Any departure from a fixed price contract in favour of cost reimbursement or target cost contracts is recommended only when the circumstances specifically require a different approach in order that the best value for money can be obtained from an employer's perspective.

The majority of the JCT forms of contract are essentially fixed price contracts, albeit that they usually contain optional clauses to allow an element of reimbursement of fluctuation costs. Optional fluctuation clauses in JCT contracts pass the risk of inflation for labour, materials, tax, levies and fuels from the contractor to the employer. However, other major areas of risk, such as management, supervision, resource availability and productivity, remain at risk to the contractor.

JCT fixed price contracts

The main JCT fixed price contracts together with the other items necessary to complete the contract documentation are set out in Table 9.1.

The Aqua Group Guide to Procurement, Tendering and Contract Administration, Second Edition.
Edited by Mark Hackett and Gary Statham.
© 2016 John Wiley & Sons, Ltd. Published 2016 by John Wiley & Sons, Ltd.

The standard building contract

The SBC in its various versions is, perhaps, the JCT's most comprehensive statement on contract conditions. The forms contain detailed conditions regulating the rights and obligations of the parties, the powers and duties of the architect/contract administrator and the quantity surveyor, and procedures appropriate for a variety of situations to be met on most projects of any size and complexity – even to the extent of covering the outbreak of war, hostilities and any act of terrorism.

The With Quantities, Approximate Quantities and Without Quantities versions all contain similar details and differ only in respect of those conditions referring to the contract bills, remeasurement (in the case of approximate quantities) and schedule of rates (where quantities are not measured).

The SBC form, derived from a long list of predecessors, has been built up from a considerable body of case law which helps clarify the manner in which the documentation is to be interpreted.

The appropriate version of the SBC form is selected and is dependent on the amount of information available at tender stage. It should be noted that with the RICS New Rules of Measurement – Detailed Measurement for Building Works (NRM2) and Co-ordinated Project Information (CPI) the choice of form used is more directly related to the information available. Bills of quantities cannot, at least in theory, be produced without the appropriate drawings and specification information being available.

Recognising the concept of contractor's design, the SBC provides for a contractor's designed portion (CDP) where the majority of the design is produced by consultants and part is to be undertaken by the contractor. If a CDP is used, the basic documentation of the SBC is reinforced by the design obligations contained in the Design and Build (DB) contract, namely the Employer's Requirements referring to that portion of the works to be designed by the contractor, the Contractor' Proposals and the Contract Design Portion Analysis.

A comprehensive set of tender and contract documentation is also available for sub-contractors. This is fully co-ordinated with the main form of contract, which ensures that the relevant conditions are fully compatible.

Provision is also made for reimbursing fluctuations in the prices of labour and materials, either by traditional means or by using the formula method when quantities are provided.

Design and build contract

In addition to the traditional fixed price contracts, there is the DB contract. This form is different from traditional forms since its concept is based on the final design being completed by the contractor in place of that ordinarily provided by the employer's consultants. The concept of the contractor providing the design requires alternative contract documents to those used in traditional procurement. These alternative documents comprise the Employer's Requirements, the Contractor's Proposals and a Contract Sum Analysis (see Chapter 13).

Part II

Table 9.1   JCT forms of contract.

Form of Contract		Other Documentation
(1) Standard Building Contract		
(a) with quantities		Drawings
		Bills of quantities (incorporating the specification)
(b) with approximate quantities		Drawings
		Bills of approximate quantities (incorporating the specification)
(c) without quantities		Drawings
		Specification
	optional	Work schedules
		Contract sum analysis/schedule of rates
with contractor's designed portion (CDP) (for optional use with quantities, approximate quantities and without quantities)	*additionally*	Employer's requirements
		Contractor's proposals
		CDP analysis
(2) Design and Build Contract		Employer's requirements
		Contractor's proposals
		Contract sum analysis
(3) Major Project Construction Contract		The requirements
		The proposals
(4) Intermediate Building Contract		
(a) with quantities		Drawings
		Bills of quantities
(b) without quantities		Drawings with or without specification
		Work schedules (priced)
	or	Specification (priced) with or without work schedules
	or	Work schedules with contract sum analysis or schedule of rates
(5) Intermediate Building Contract With Contractor's Design	*additionally*	As the intermediate building contract
		Employer's requirements
		Contractor's proposals
		CDP analysis
(6) Minor Works Building Contract		Drawings
		Specification
		Work schedules
		Schedule of rates (priced)
	or	Drawings
		Specification (priced)
		Work schedules
	or	Drawings
		Specification
		Work schedules (priced)
(7) Minor Works Building Contract With Contractor's Design	*additionally*	As the Minor Works Building Contract
		Employer's requirements

Major project construction contract

The Major Project Construction Contract (MP) was originally drafted under the guise of the Major Project Form in 2003. Guidance to the MP advises that this form is for employers who have in-house contractual procedures and regularly procure large-scale construction projects. The contractual provisions are therefore somewhat shorter than those of the SBC. The form is similar to that of the DB contract and utilises requirements which are, in principle, the same as Employer's Requirements in the DB contract. However, the form does not preclude the use of traditional procurement. The requirements may simply comprise drawings and bills of quantities but this is unlikely as those who are envisaged as using the contract are also likely to seek to transfer the design risk to the contractor.

The MP is sensible in its approach to the responsibility of sub-contractors. If a sub-contractor of the employer's choice is required to undertake part of the project, the contractor assumes full responsibility for the sub-contractor without systematically jumping through the hoops required under the naming provisions of the Intermediate Building Contract (IC) or the nomination procedure formerly practised under the SBC's predecessor, the JCT 98. In allowing the employer to name a sub-contractor and ensure that the contractor is responsible for the sub-contractor, the requirements simply need to state so.

As with the employer requiring the contractor to be responsible for subcontractors, the form also envisages that the employer's consultants, who were appointed to draft the requirements, might at a suitable time into the project, be novated to the contractor. However, the contract does not set out the terms on which novation must take place, but merely refers to a model form to be referenced in the requirements. Therefore, one would expect that a model form would be included as part of the requirements so that the contractor is aware of the terms and conditions stated therein and to ensure that he has the opportunity to price the risk before the contract is entered into.

The MP has a practical approach to completing a project for the mutual benefit of both parties. The form provides for:

- Accelerating the project to practical completion before the completion date.
- Bonus payments should practical completion occur before the completion date.
- Cost savings and value improvements for which a benefit may arise in the form of:
  - a reduction in the cost of the project;
  - a reduction in the life cycle costs associated with the project; and
  - any other financial saving to the employer.
- Novation of the design team.

The form is drafted in the manner that encourages the contractor to suggest amendments to the requirements and/or proposals which would result in a financial benefit to the employer.

If there is a disadvantage in using this form over the SBC or DB it is that, unless the employer is experienced in procuring projects and can provide much of the input as to the requirements, the additional cost of professional fees ordinarily incurred is likely to be above that of other JCT forms. In contrast to the SBC and

DB forms, the MP is more facilitative in its approach than comprehensive and is for use by employers with experience and precedent to draw upon.

Intermediate building contract

The IC is produced in two formats. The first format is based on the premise that all of the design is completed by the employer's design team and the second envisages that the contractor is required to complete part of the design. Essentially the main difference between the two documents is that the Intermediate Building Contract with Contractor's Design (ICD) allows for contractor's design by containing a CDP in much the same way as the option within the SBC.

To give a degree of flexibility, the form contains a number of alternative clauses in the recitals and contract particulars which, depending upon the particular circumstances of the project, are selected (deleting those not required) to produce a with quantities or without quantities format. The form contains conditions that are shorter and somewhat less detailed than those of the SBC variants but more detailed than those of the Minor Works (MW) Contract. The IC Contract Guide suggests that the form is primarily intended for use where proposed works are:

- of simple content involving the normal, recognised basic trades and skills of the industry;
- without any building service installations of a complex nature, or other complex specialist work;
- designed by or on behalf of the employer (although ICD caters for an element of contractor's design); and
- defined adequately in terms of quality and quantity in the bills of quantities, specifications or work schedules.

The provisions for the reimbursement of fluctuations in the prices of labour and materials are similar to the SBC, albeit shorter, allowing for the traditional or formula methods as preferred (the formula method only being available when quantities are provided).

The IC provides employers with the opportunity to systematically select individual sub-contractors to undertake the works. These are identified within the contract as named sub-contractors. The advantage of naming sub-contractors is that they can be selected in advance of the main contractor, should specialist work be required; once a sub-contract is entered into, with the exception of issues arising from design liability, the named sub-contractor effectively becomes a domestic sub-contractor and the main contractor assumes total responsibility for the work (subject to various special conditions if the sub-contract has to be determined). Special tender and sub-contract forms are provided to administer this process to ensure compatibility with the main contract. It is important that this procedure is carried out in advance of the signing of the main contract as any sub-contract must be executed within 21 days of executing the main contract.

In this form there is authority for the employer to appoint a clerk of works. The clerk of works has only one duty and that is to act solely as inspector on behalf of the employer and under the directions of the architect/contract administrator.

Minor works building contract

The Minor Works Building Contract (MW) is one of the simpler JCT forms available and, like the IC, is available in two formats: one in which all design work is completed by consultants on behalf of the employer and the other in which the contractor undertakes to complete part of the design (MWD). The form is drafted for use with drawings and/or specifications and/or a work schedule, but no provision is made for the use of bills of quantities. MW guidance suggests that it is appropriate to use the form where:

- the work involved is simple in character;
- the work is designed on behalf of the employer (except the contractor's designed portion in MWD); and
- a contract administrator is appointed by the employer to administer the conditions.

Although minor works might not be considered to extend over a long period of time, provision is made for fluctuations in the prices of labour and materials due to contribution, levy and tax changes.

In situations whereby a specialist sub-contractor is required to undertake the design of part of the works, the Minor Works Sub-Contract with Contractor's Design (MWSub/D) allows the design to form part of the contractor's design under the contractor's design portion of the MWD. The contractor does, however, remain wholly responsible for any works undertaken by a sub-contractor.

However, with regard to the provisions of other JCT forms, there are a number of significant omissions from the MW contract. These omissions comprise:

- any provisions for insuring the employer's liability for damage to property other than the works, or for all-risks cover other than the works;
- any provision for the ownership of materials on site to pass to the employer on payment; and
- any specific provision for opening-up and testing.

In respect of the first point above, there are provisions in the contract for the contractor to indemnify the employer against any expense, liability, loss, claim or proceedings for damage to any property.

As with the IC, or any other contract for that matter, any limitations could be overcome by drafting special clauses stipulating the employer's particular requirements. However, there seems little point in doing so when there are more comprehensive forms available covering the omissions in this very simple form.

Other fixed price contracts available

It is worthy of note that the JCT publishes further contracts. These comprise:

- Constructing Excellence Contract (CE) – such is the flexibility of this contract, it could also be a target cost contract;

- Repair and Maintenance Contract (Commercial) (RM);
- Measured Term Contract (MTC);
- Building contract for a home owner/occupier who has not appointed a consultant to oversee the work (HO/B);
- Building contract for a home owner/occupier who has appointed a consultant to oversee the work (HO/C); and
- Home Repair and Maintenance Contract (HO/RM).

It is, however, beyond the scope of this book to cover each and every contract published by the JCT. Further details in respect of these contracts can be found in the guidance document entitled *Deciding on the appropriate JCT contract 2011* published by the JCT.

Advantages and disadvantages of fixed price contracts

The following points may be found useful in deciding whether to recommend to a client the use of a fixed price over a cost reimbursement contract.

Advantages

- Depending on which elements of the fluctuations clauses are deleted, the employer avoids the risk of fluctuations in cost due to management deficiencies, shortage of resources, reduced productivity and, to some extent, inflation.
- Unless approximate quantities are provided the contract sum will define the employer's financial commitment, subject to contingency and any other provisional allowances and/or any variations in the brief.
- There is an in-built incentive for the contractor to manage the work effectively and to complete the work as quickly as possible in order to maximise his profit.

Disadvantages

- The premium paid to the contractor for taking the risk is paid irrespective of whether or not the risk materialises.
- Time and production cost savings and improved productivity deriving from the learning curve are lost to the client; only the contractor benefits. This may be particularly significant on repetitive jobs.

The choice between the various fixed price contracts available will depend on the other documentation available and the scope and complexity of the works.

Chapter 10
Cost Reimbursement Contracts

In Chapter 8, cost reimbursement was defined in terms of payment on the basis of the actual cost incurred by the contractor, irrespective of any estimate which may have been calculated for budgeting purposes or submitted as part of a tender. A cost reimbursement contract is therefore one in which the contractor is reimbursed actual cost plus a fee to cover overheads and profit.

By their very nature, cost reimbursement contracts cannot have a finite sum either at contract stage or when a contract is signed. The sum due will only be ascertained at completion when the final account is settled.

Since the contractor is reimbursed the actual cost, it will be seen that, with this type of contract, the employer carries all the risk – inflation, management efficiency, effective supervision, resource availability and productivity. The contractor does take on a contractual obligation to carry out the work as economically as possible, with regard to the nature of the works, the prices of materials and goods and the rates of wages current at the time that the work is carried out, together with other relevant circumstances. It is nevertheless important that employers and their consultants have confidence in the competence of the contractor and that some way of controlling the contractor's method of operation is included in the contract documentation.

The fee

The fee paid to the contractor can be a percentage figure (applied to the prime cost) or a fixed fee (a lump sum figure). A fixed fee has the advantage of providing an incentive for contractors to work efficiently (to maximise their return). A percentage fee lacks this incentive and is sometimes considered to be an open cheque for contractors – the more they spend, the higher their fee – and is very much akin

The Aqua Group Guide to Procurement, Tendering and Contract Administration, Second Edition.
Edited by Mark Hackett and Gary Statham.
© 2016 John Wiley & Sons, Ltd. Published 2016 by John Wiley & Sons, Ltd.

to the concept of dayworks. Although this may be considered an extreme view, it does serve to highlight the danger of not providing an incentive for contractors to work efficiently. However, if the scope of the work is not sufficiently defined, a fixed fee may not be acceptable to any contractor and a percentage may have to be agreed in any event.

A refinement of the fixed fee method is to incorporate a provision for the fee to be varied as the final estimate of prime cost varies in relation to the original estimate of prime cost. This helps in dealing with variations.

As a further incentive to efficiency, it is possible, with either a percentage or fixed fee basis, to incorporate a provision that the fee is only adjusted if the final estimated prime cost varies from the original estimated prime cost by more than 10%, or some other agreed percentage. This adjustment could apply to both upward and downward movement.

The prime cost building contract

A fairly typical view held in the construction industry is that fixed price contracts may be considered to be the norm and that cost reimbursement contracts should only be used on those projects where specific conditions render them more suitable, for example, emergency repairs, investigative works and such other cases where it is not possible to identify in advance the scope of the work required to be done.

Nevertheless the advent of partnering arrangements has led to an increase in the use of this form of contract. If it is properly administered by both the contractor and the architect/contract administrator, the employer may benefit from not paying for risk items that do not materialise. It might also be said that this approach prevents the situation of a well managed and effective contractor from suffering a loss on a poorly defined and problematic project.

Characteristics of the form

The main feature distinguishing this form from that of fixed price contracts is the agreement for payment to the contractor. Payment is based on the prime cost of the work, as defined in the documentation, and a fixed or percentage fee.

The form contains detailed conditions regulating the rights and obligations of the parties, the powers and duties of the architect/contract administrator and the quantity surveyor and procedures appropriate for the administration of the work.

By its very essence, a cost reimbursement contract does not make any provision for reimbursement fluctuations; these are dealt with automatically as invoices and time sheets are priced at rates current at the time the work is carried out.

The PCC provides definition of Prime Cost and divides the costs to be incurred into six categories in order that conditions can be applied to the costs incurred in respect of each category. The definition provides that the Prime Cost shall comprise the sum of the following VAT-exclusive costs insofar as they are reasonably and properly incurred in accordance with the Contract:

  (i)   contractor's staff on site;
  (ii)   contractor's direct workforce;
  (iii) materials and goods provided by the contractor;
  (iv) plant, services and consumables stores;
  (v)   sundry costs; and
  (vi) sub-contract work.

In calculating the Prime Cost:

- no amount of loss and/or expense shall be included;
- no amount of costs and expenses arising from the contractor's right to suspend performance shall be included;
- there shall be no inclusion of any costs disallowed because the contractor has, in the opinion of the contract administrator, not complied with the obligation to carry out works as economically as possible with regard to all of the circumstances;
- there shall be no inclusion of any costs disallowed because a greater number of persons were on site than reasonably required to carry out and complete the works;
- there shall be deducted any discounts received by the contractor in relation to carrying out the works and/or any payments to or credits received (for materials etc.) by the contractor in the course of carrying out the works;
- for any period of default in the completion of the works or a section, payments shall be limited to amounts calculated at the rates applicable immediately before the completions date, or (if less) their actual amount.

The PCC also provides the use of specific rates and/or lumps sums to be included as part of the payment mechanism in lieu of the prime cost for each of the six categories identified above. Thus, in using the PCC, an employer is able to fix as much of the project costs as the particular circumstances permit.

Advantages and disadvantages of cost reimbursement contracts

Since fixed price contracts are considered suitable for most circumstances, there should be cogent reasons for recommending to a client the use of a cost reimbursement contract. Consideration should be given to the advantages and disadvantages of such a contract when evaluating whether or not the use would be to a client's advantage in any particular situation.

Advantages

- Work can be commenced on site immediately, provided that sufficient specification information is available.
- It is often the only way of carrying out investigation work.

Part II

- It is ideal for coping with emergencies such as making buildings safe after fire damage.
- Production cost savings and improved productivity can derive from the learning curve resulting from the repetitive nature of jobs, to the benefit of the client.
- The client pays no premium to cover risk – he pays only the actual cost of whatever is deemed to be necessary if and when those risks actually materialise.

Disadvantages

- The client's ultimate commitment is not known.
- There is little incentive to the contractor to employ his resources efficiently (but see section above on the fee).
- The client carries nearly all the risk on the contract.
- Checking the prime cost can be a complicated and expensive operation.

Budget and cost control

The procedure for creating a budget and controlling the cost in the early feasibility and design stages is similar to that for a normal fixed price contract. An early budget estimate should be prepared and there is no reason why cost planning should not go ahead normally. Furthermore, cost checking during the evolution of the design can also be carried out as with other types of contract.

It is important that the documents showing the estimated cost of the project should be as detailed as possible so that the work is clearly defined, and that the basis on which the fee is calculated is established, whether it be a fixed fee or a percentage fee. It is also necessary that the estimated prime cost should be divided between the work to be carried out by the contractor's own labour and the work to be carried out by sub-contractors.

If it is decided that the fixed fee should be settled in competition and the contractor selected in that way, the detailed estimate must be sent to the tendering contractors as one of the tender documents.

Administering the contract

The procedures for operating fixed price contracts (the norm) are dealt with in some detail in Part IV, Contract Administration. This chapter considers only those procedures relating to cost reimbursement contracts which differ from the norm.

It is not normally advisable to change clauses in standard forms of contract, but the Prime Cost Building Contract does contain a good deal of procedural matter which it is perfectly reasonable to vary in order that the most efficient method of working can be established on a particular project. For example, if joinery is likely to be supplied by the contractor's own workshop, it may be desirable that the calculation of prime cost for that work should be on a different basis to the prime cost of work on site. It is also permissible to consider the incidence of small tools, perhaps even consumables, and to judge whether it is likely that fewer disputes will arise if some of these items are included in the fee rather than the prime cost.

There are other matters of procedure in which the quantity surveyor will be particularly interested. For instance, for cost control purposes, the contractor may be required to send in time sheets and invoices at regular intervals for checking. Often a procedure for checking deliveries and delivery notes is necessary. It is desirable to ensure that the contract covers the procedures required.

Procedure for keeping prime costs

The definition of prime cost is crucial as generally what is not specifically included in the definition is automatically deemed to be included in the fee. With small contractors, who might be unused to working by this method, it is wise to point this out in some detail before the contract is signed so that there is complete understanding on both sides as to what is supposed to be included in the fee. This is the best way to avoid niggling disputes,

The quantity surveyor should agree with the contractor a working method of keeping the prime cost. Contractors are sometimes inclined to take the line that, as they have been chosen to carry out work on this basis, they should have a free hand and their method of keeping the prime cost should be accepted. However, a proper system of checking the prime cost is essential and does not indicate mistrust of the contractor: it is merely a prudent method of doing business that is only right and proper on accountability grounds. Furthermore, the contractor's internal method of keeping the prime cost may not, and in fact probably will not, coincide exactly with the definition of prime cost in the contract. The quantity surveyor will, however, be well advised to go along as far as possible with the contractor's normal costing system as this is certainly the most likely to produce error-free results.

There seems to be no reason why the quantity surveyor should not receive a copy of the weekly wage-sheet, particularly where this is prepared for the contractor in any case. If some different method is used, this must be discussed with the contractor. Quantity surveyors should not object to using another method, provided they get the same information as the contractor with regard to wages paid.

Contractor's site staff and direct workforce

It is perhaps worthwhile at an early stage, just after the contractor is selected, to agree on who is going to be the contractors' site management staff and the direct workforce which are to be paid as part of the prime cost and which people of a supervisory capacity are covered by the management fee (if any).

A further fundamental point arising with cost reimbursement contracts is that, as the clients are taking the risk, they should have some control over the way the work is carried out. For this reason it is desirable that the extent of normal overtime to be worked is agreed in advance and that the contractor should then seek approval before working any further overtime. Thought should be given to all such matters, according to the type of project, and decisions made as to what should be incorporated in the documents for tendering.

Materials

A system must be worked out for the acquisition of materials. Generally speaking, competitive quotations should be obtained but it must be recognised that there may be occasions when materials are wanted so quickly that this is not possible. In such a case the quantity surveyor has to take a reasonable view and provided that the materials bought are not above market price, there is no reason why they should not be included and paid for. Even where competitive quotations can be obtained, the quality of service likely to be given to the contractor should also be considered. It may be that the lowest quotation is not the most advantageous. Such instances are, however, likely to be rare, particularly so where quotations are only sought from reputable suppliers.

Some system must be worked out to ensure that materials invoiced match up with materials delivered. This can be done by means of delivery tickets, and frequently contractors work such a system on their own fixed price contracts. An overall check on any particular material can be made by an assessment from the estimated prime cost. In many cases a physical check on site can easily be carried out.

Plant

The conditions of contract should state how plant is to be charged. The two main plant hire schedules in current use are:

- The BCIS Schedule of Basic Plant Charges.
- The Civil Engineering Contractors Association Schedules of Dayworks.

Although both of these schedules are intended for use in connection with day-works under a contract, they can, with suitable adjustment, be used for cost reimbursement contracts. Generally, however, it is assumed that if the plant is wanted for a long period on site, it may be cheaper to buy it outright. It may be desirable therefore that some limit upon the period of hire of plant should be laid down in the contract. Sometimes this is expressed in terms of not paying more in hire than, say, 80% of the capital cost. At the end of the contract, purchased plant can be sold or retained by the contractor, with an appropriate credit being included in the prime cost calculation.

Plant hire will invariably bring with it the question of consumable stores and discussion should also take place early on how these will be treated.

Credits

A system should be established for allowing credits for surplus materials and also credits for salvage, such things as old lead taken in from existing buildings.

Similarly, material supplied by the client must be taken into account when agreeing on the management fee. The sale of purchased plant has already been referred to above.

Sub-letting

On all cost reimbursement contracts, there will be a proportion of work which is sub-let on a measured basis. This may be to a sub-contractor, named or otherwise, or possibly to a labour-only sub-contractor. Provided the arrangement is economical and competitive, there is no reason why it should not be adopted. However, if the proportion of work being sub-let rises higher than that estimated when the fee was originally fixed, there may be a case for a variation of the fee. This will have to be established before the contract is signed and, for this reason, it is desirable that any contractor's intentions regarding sub-letting on this basis should be established before they are finally selected.

Defective work

Most prime cost contracts have a clause indicating that the contractor should make good defects at his own expense. This can raise problems. If the defects are made good after practical completion there is no difficulty in separating the costs involved, but if they are made good during the progress of the work then some specific separation of the cost must be made. A kind of negative daywork charge is needed; this is very difficult to apply in practice.

Cost control

Valuations for cost reimbursement contracts will normally be done on the basis of the labour paid and the invoices submitted by the contractor each month. The retention may be a percentage of the value of the work done or a proportion, say 25%, of the fixed fee. There is always a considerable interval between work being carried out and invoices being submitted and therefore, for cost control purposes, a specific reconciliation between the actual prime cost and the estimated prime cost must be made.

To do this properly one has to go back to the detailed estimate of prime cost used for tendering and establish what proportion of that work is done. Against this, one sets the actual prime cost, plus the labour and material used but not yet invoiced. The difference between the two will show the extent to which the project is either saving on (or exceeding) the estimated prime cost. From this reconciliation, estimates of the final cost of the project can be made. Obviously, in taking the original estimate of prime cost, allowance has to be made for any variations. Although cost control with cost reimbursement contracts is more difficult than with fixed price contracts, it must be carried out and the outcome should be reasonably satisfactory.

Final account

The final account will, of course, be the total of the actual prime cost plus the management fee. It may be that during interim valuations various items are put into a suspense account pending settlement as to whether or not they are a proper charge against prime cost. If possible, items in the suspense account should be settled and either included or rejected as soon as possible. The procedure for the final certificate, maintenance and handing over of drawings and so on in relation to the building is similar to that of normal fixed price contracts.

Chapter 11
Target Cost Contracts

Chapter 8 examined the relationship between fixed price and cost reimbursement contracts and established the position of the target cost contract as a half-way house between them. Target cost contracts are a refinement of ordinary cost reimbursement contracts and introduce an element of incentive for the contractor to operate efficiently by transferring a proportion of the risk to the contractor. These are sometimes referred to as 'gain-share pain-share' contracts.

The essential feature of the target cost contract is that the difference between the actual cost and the estimated cost is split in some way between the contractor and the employer. The philosophy behind this is that if the contractor cannot complete the work for the estimated cost, it is not right for the employer to pay the whole actual cost, as at least some of the increase may be due to the inefficiency of the contractor rather than to inaccurate estimating. On the other hand, and equally importantly, if the contractor completes the work at a lower cost than that estimated, it may be assumed that some of the decrease is due to his own efficient management and, therefore, some of the gain should go to him. In this way the target cost contract gives the contractor a built-in incentive to manage the work as efficiently as possible.

It also, of course, provides the contractor with an incentive to increase the estimated price as much as possible in the first place. It is therefore essential that employers take steps to ensure that their interests are safeguarded by employing expert advice in evaluating the estimated price. This may be negotiated with the contractor before work is started or can be established by competition as part of the original tendering process.

Target cost contracts, however, should not be entered into lightly. They are expensive to manage, involving as they do both accurate measurement and careful costing on the employer's behalf. This is no doubt one reason why target cost contracts are not as prevalent as other forms but it is important to be aware of the availability of the system should the need arise.

The Aqua Group Guide to Procurement, Tendering and Contract Administration, Second Edition.
Edited by Mark Hackett and Gary Statham.
© 2016 John Wiley & Sons, Ltd. Published 2016 by John Wiley & Sons, Ltd.

	£
Target (i.e. estimate total prime cost including allowance for head office overheads and profit)	100,000
Actual prime cost plus overheads, etc.	90,000
Therefore saving on target	10,000
If saving is split on a 50:50 basis, payment to contractor becomes	95,000

Example 11.1   Target cost contract with saving.

Target (as before)	100,000
Actual prime cost plus overheads, etc.	110,000
Therefore deficit between actual cost and target	10,000
If deficit is split on a 50:50 basis, payment to contractor becomes	105,000

Example 11.2   Target cost contract with overspend.

In a target cost contract, payment is made partly on an estimated fixed price basis and partly on the actual prime cost. This principle can best be understood by looking at two simple examples, one showing a saving on the original target and the other showing extra:

There are various points to consider in studying these examples:

- The figures above have been deliberately made simple: in practice the target must be revised to account for all the variations. It will, therefore, be the revised target which is compared with the actual cost to establish the saving or extra.
- The split can be in any proportions previously agreed. Thus, if it is desired that the contractor should take more of the risk, the proportions will be agreed so that the employer pays a figure nearer to the revised target and further from the actual prime cost.
- Various methods can be used to achieve a more sophisticated calculation by splitting the work between that of contractors themselves and that of their own and named sub-contractors. In addition, their head office overheads and profit can be covered by either a fixed fee or a percentage fee. In a case of a fixed fee, it would only alter if the revised target varied from the original.
- The target can be obtained in various ways. The importance of reasonable accuracy has already been stressed. The best way, undoubtedly, is by means of a full and accurate description of the whole, properly priced out and agreed between the contractor's and employer's quantity surveyors. There may, however, be circumstances when a bill of approximate quantities will suffice. For specialist work, a method of estimating in common use by the particular trade, such as a labour and material bill, may be appropriate.

In the simple examples given, prime cost is assumed to be prime cost to the employer in accordance with the detailed terms of the contract defining prime

	Target	Actual	
		A	B
	£	£	£
(1) Prime cost of labour charges, clearly defining rates, allowances, fares, etc.	20,000	22,000	18,000
(2) Prime cost of materials and consumable stores, making allowance for discounts	25,000	26,000	24,000
(3) Sub-contractors work	20,000	21,000	19,000
(4) Site management (fixed)	12,000	12,000	12,000
(5) Plant, sheds, transport, etc. (fixed)	9000		9000
		9000	
(6) Office overheads, supervision and insurance (fixed)	8000		8000
		8000	
(7) Profit or fee (fixed)	6000		6000
		6000	
Column A – reduction in profit in respect of items (1) and (2) 50% of £3000	100,000	104,000	96,000
		1500	
Cost to employer			
Column B – additional profit in respect of items		102,500	
(1) and (2) 50% of £3000			1500
Cost to employer			97,500

Example 11.3   Target cost contract with cost saving and overspend.

cost. For practical purposes, however, it may well be appropriate to negotiate some of the items on a fixed price basis. There are many possible variations and Example 11.3 illustrates one possible solution where three elements of the work involving site management (4), plant (5) and overheads (6) are on a fixed price basis, while being incorporated into an overall target cost contract. Sub-contractors (3) are included on a full cost reimbursement basis and, for the sake of simplicity, the profit or fee (7) is taken as fixed. This leaves the labour (1) and materials (2) elements (amounting to 45% of the total) subject to the target cost calculation.

Guaranteed maximum price contracts

This is a type of target cost contract where the agreed guaranteed maximum price cannot be exceeded.

Any saving on the guaranteed maximum price would be shared between the contractor and the employer in a similar fashion to column B in Example 11.3.

Guaranteed maximum price contracts are very difficult to set up and administer for two fundamental reasons. First, it is difficult to agree on a realistic maximum price, as a contractor will inevitably try to set the figure too high, making it meaningless, and secondly, all variations will require the maximum price to be reassessed.

Competition

Very often the target will be negotiated between quantity surveyors on both sides. However, it is possible for the target to be the subject of a competition between contractors. This particularly applies in civil engineering works. In the latter case the bills of quantities will form the tendering document on which the target is based. Normally the contractor submitting the lowest target is selected.

From then on, the procedure is the same as with a negotiated target. The work is measured and valued on the basis of the bills, any variations being taken into account. The contractor is paid on the basis of the actual prime cost plus the relevant fee, and the difference between that figure and the measured account is shared between the contractor and the client on the lines indicated earlier.

Contract

The CE contract contains alternative payment regimes for Target Cost (including the possibility of a Guaranteed Maximum Cost) or lump sum. It is essential to understand the term Allowable Cost. The Allowable Cost is the benchmark to which the Target Cost is compared for the purpose of the 'gain-share pain-share'. It is possible that three situations could arise in the event that the Target Cost option is used:

1. Insofar as the Target Cost for completing the services (works) is below the Allowable Cost, the savings are shared between the Purchaser (employer) and Supplier (contractor) in the relevant proportions agreed in the contract.
2. Insofar as the Allowable Cost for completing the services is greater that the Target Cost but is less or equal to the Guaranteed Maximum Cost (if any), the Supplier shall be entitled to receive only the relevant proportion agreed in the contract. This effectively means that the Supplier is likely to contribute to the construction cost of the project.
3. Insofar as the Allowable Cost for completing the services exceeds the relevant Guaranteed Maximum Cost, the amount in excess shall be borne by the supplier. The amount between the Target Cost and Guaranteed Maximum Cost (if any), shall be shared in accordance with item two above.

It is not controversial to say that any suggested changes to the Allowable Cost might provide fertile grounds for dispute between the parties. On the one hand, there is an incentive to receive monies for savings and, on the other, there is the risk of being held fully responsible.

Advantages and disadvantages

The advantages and disadvantages of target cost contracts are very much in line with those of cost reimbursement contracts identified in Chapter 10. The only difference, and it is an important one, is that target cost contracts have the further advantage of including an incentive, encouraging the contractor to operate as efficiently as possible. There is also an advantage in the flexibility of the system. The risk can be apportioned in different degrees depending on the circumstances of each party and their ability to take that risk. It need never be a black and white solution; all shades of grey are attainable according to the merits of each case.

The incorporation of a guaranteed maximum price in a target cost contract introduces additional factors for evaluation. The advantage to the employer of the price ceiling has to be set against the disadvantages of a higher fee – for it must be appreciated that a higher premium will be charged for the additional risk taken by the contractor.

Use

The likely uses of target cost contracts have already been reviewed in Chapter 8, when considering the general use of cost reimbursement contracts, but, as indicated above, they do have the additional advantage of providing an incentive for the contractor to improve efficiency.

Part II

Chapter 12
Management and Construction Management Contracts

Management and construction management contracting are forms of contractual arrangement whereby the contractor is paid a fee to manage the building of a project on behalf of the employer. They are therefore contracts to manage, procure and supervise rather than contracts to build.

Under management and construction management agreements, the contractor becomes a member of the employer's team – effectively the construction consultant – and it is from this factor that one of the main advantages of the arrangements is derived – the contractor working in harmony with and as part of the project team.

Whilst management and construction management contracts are very similar in many respects, there is one essential distinguishing characteristic which is fundamental to the understanding of the two systems. This involves the contractual arrangements between the parties. In management contracting, the works contractors, or package contractors, are in contract with the management contractor; in construction management, they are in contract with the employer. A diagrammatic representation of this difference is shown in Figure 12.1. It will be noted that, apart from this contractual link, the lines of communication are identical. This difference will be referred to again later, but the two systems can be evaluated together. For convenience, the term contractor/manager is used to denote the management contractor or construction manager.

Payment and cost control

To understand the way a management or construction management contract works, it is necessary to show how payment is made.

The building work is split into sub-contracts, generally referred to as packages. These are normally let in competition in the usual way, although some may be

The Aqua Group Guide to Procurement, Tendering and Contract Administration, Second Edition.
Edited by Mark Hackett and Gary Statham.
© 2016 John Wiley & Sons, Ltd. Published 2016 by John Wiley & Sons, Ltd.

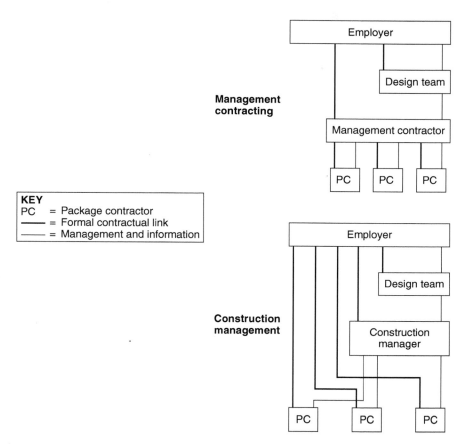

Figure 12.1 Differing contractual arrangements.

negotiated or let on a cost reimbursement basis. All discounts usually revert to the client. It should be possible to arrange the various sub-contracts so that all building work is covered, but in practice a small site gang will sometimes be required to unload and help with the movement of materials around the site and general cleaning after trades. Apart from this it is unusual for management contractors to carry out any of the building work themselves, although there may be other companies within the management contractor's group wishing to tender.

The contractor/manager provides the site management team and such preliminary items as offices, canteen, hoardings and so on, which are usually reimbursed at cost. On some projects, however, the cost of these forms part of his tender and is a fixed lump sum. In either case, a schedule forming part of the contract clearly defines what is required.

In addition, the contractor/manager will be paid a fee. This fee is in respect of overheads (i.e. head office charges) and profit and this is usually expressed as a percentage of the final prime cost of the works. It is therefore important to distinguish the site management team from the head office staff and facilities involved. Alternatively, the fee may be based on the cost plan and not be subject to change. Thus if the cost goes up, for no good reason, the effective percentage for overheads and profit on the actual cost decreases.

A cost plan will have been prepared prior to the appointment of the contractor/manager. As soon as the contractor/manager is appointed, he must liaise with the quantity surveyor to prepare an estimate of the prime cost showing the estimated value of all the sub-contract packages, site management and other costs. This is then agreed by the client and architect as reflecting the required level of specification and becomes the estimate of prime cost for the project. Valuations and certificates for payment are prepared in the usual way. Cost reporting is carried out on a package by package basis against the estimated prime cost.

Some employers require the contractor/manager to be bound by a guaranteed maximum price (or GMP), this being a sum above which any expenditure will not be reimbursed. This, however, cannot be achieved until the design and specification of the building are fairly advanced and therefore, given the nature of these contracts, cannot usually be stated in the original contract documentation.

Selection and appointment of the contractor

Selection of the contractor/manager can either be on a negotiated basis or, as is more usually the case, in competition. Negotiation is easier than for a fixed price contract as the payment for most of the work relates to sub-contracts, which will be dealt with separately after selection of the contractor/manager. Negotiation would be on the basis of the management fee together with the contractor/manager's estimate of site management requirements.

If the contractor/manager is to be selected in competition, tender documentation should be produced inviting the submission of proposals. This documentation will advise the tenderers of everything known about the project which, given the early stage at which contractor/managers are often appointed, may not be very much. It will, however, generally include:

- general arrangement drawings;
- known specification information;
- the expected contract value;
- details of any key dates; and
- details of the contract document.

The contractor/manager's submission would normally include the following:

- the management fee;
- an estimate (or tender if required) of the site management and preliminary costs;
- any comments on the estimated cost of the work, given the other information available;
- a method statement giving outline proposals for carrying out the work, including a draft list of work packages (sub-contracts);
- a draft contract programme;
- details of the proposed management team; and
- a proposed typical sub-contract form.

Clearly, selection will be made largely on the credibility of the contractor/ manager to provide the building on time and within the budget since a difference of, say, 0.5% in the fee may not be significant in the context of some of the problems which could occur on a major development. With this in mind, the contractors/managers will be anxious to include in their submissions details of their track record of successful outcomes by reference to past projects.

Some or all of the contractors/managers are generally invited to attend an interview with the employer and the design team principals, after which an appointment is made. This is usually on the basis of a Letter of Intent, which may be followed up by a Pre-Construction Agreement. The purpose of the latter agreement is to define the responsibilities of the parties and the formula for reimbursement (usually a time charge for staff and actual expenditure) in the event that the project is aborted before commencement on site.

Contract conditions

All the management contracting and construction management firms and many clients, architects and quantity surveyors have for some years now had their own forms of contract. However, the JCT publishes standard forms for construction management and management contracting. These are the Construction Management Contract and the Management Building Contract respectively. Both contracts comprise a series of sub-agreements.

The Construction Management Contract comprises:

- Construction Management Appointment
- Construction Management Trade Contract   CM/A
- Construction Management Guide   CM/TC
- Construction Manager Collateral Warranty for a Funder   CM/G
- Construction Manager Collateral Warranty for a   CMWa/F
  Purchaser or Tenant   CMWa/P&T
- Trade Contractor Collateral Warranty for a Funder
- Trade Contractor Collateral Warranty for a Purchaser or   TCWa/F
  Tenant   TCWa/P&T

The Management Building Contract comprises:
- Management Building Contract
- Management Works Contract Agreement   MC
- Management Works Contract Conditions   MCWC/A
- Management Works Contractor/Employer Agreement   MCWC/C
- Management Building Contract Guide   MCWC/E
- Management Contractor Collateral Warranty for a Funder   MC/G
- Management Contractor Collateral Warranty for a   MCWa/F
  Purchaser or Tenant   MCWa/P&T
- Works Contractor Collateral Warranty for a Funder
- Works Contractor Collateral Warranty for a Purchaser or   WCWa/F
  Tenant   WCWa/P&T

Part II

Contract administration

In management contracting and construction management it is often the case that the whole of the management team (i.e. the design team, the quantity surveyor and the contractor/manager) is based or is at least represented at senior level on site. This facilitates day-to-day communication and problem solving, consistent with the objectives of fast and flexible building. Most employers will also find it necessary to designate or appoint their own project manager, who becomes the single point of contact for the construction team.

One of the management team's first tasks is to draw up a detailed programme and procurement schedule. These show the dates by which design information is required by the quantity surveyor for the production of tender documents for each package. A list of possible works contractors is also compiled. These companies are then pre-qualified to ensure they are suitable, interested and financially able and a number, usually four to six, are shortlisted to go on the package tender list. Tenders are invited by the contractor/manager and will include full preliminaries, a works-contract programme and other details such as safety and quality control procedures. Some contractors/managers also hold mid-tender interviews to make sure that the tenderers understand their part in the whole project. Any significant points arising at these meetings are, of course, circulated to all tenderers.

When the tenders are received, they are promptly reviewed by all members of the team and one or more of the tenderers may be called to a post-tender interview, especially if any queries arise. A team recommendation is then made to the employer and the package contractor is appointed.

Drawings and architect's/contract administrator's instructions are issued in the normal way. Contractor's/manager's instructions are issued to the package contractors within the authority of the architect's instructions and these form the basis of the package final account. Valuations are also carried out in the normal way with package contractors' applications being checked and audited by the quantity surveyor.

Periodical reports are made by the team to the employer, reporting on progress, on any problems being encountered or foreseen, any corrective action being taken and, of course, on the anticipated final account.

Professional advisers

Management or construction management contracts may include or exclude design work but the employer will need independent professional advice, whether or not the design is included. In the case of the employer's architect and engineers, the extent of their involvement will, of course, vary according to where the design is placed. The quantity surveyor's role is very similar to normal building contracts, but the emphasis is more on cost control and auditing of expenditure by the contractor/manager.

Advantages and disadvantages

Advantages

Among the advantages claimed for management contracting and construction management are the following:

- *Harmony, not confrontation* – Traditional contracting has been described by some as adversarial. In management contracting and construction management, however, the sole objective of all parties is to produce the required building on time and on budget and the contractor/manager is paid a fee for this service in a way similar to the other members of the client's team. The process should, therefore, be more harmonious.
- *Earlier start on site* – Management contracting and construction management contracts allow the project to be on site earlier than by other methods. Figure 12.2 shows that because sub-contract tenders are invited and let whilst design is still progressing, there can be more time for each of these procedures to take place whilst allowing a completion date which is earlier than by a more traditional method.
- *Design flexibility* – Management contracting and construction management contracts allow the employer and designers total flexibility to develop the scheme, which enables each and every proposed design to be assessed in terms of the three factors affecting it, that is, time, cost and quality. It is usual on a large project for the design team to be based on site.
- *Buildability* – Early involvement of contractors/managers means that they can advise on the suitability of proposed materials and methods with regard to time, market availability and so on.
- *Early completion* – Whilst the contract states a completion date, the client can, by agreement, require the works to be completed earlier, although this may attract payments for acceleration of the various work packages affected.
- *Less pricing risk* – As packages are tendered closer to the time of the work being executed on site, tenderers do not have to include a premium for pricing risk.

Disadvantages

Amongst the disadvantages are the following:

- *Possible escalation of cost* – Some people express concern over cost because the contractor can be appointed whilst the design is still conceptual, but in fact almost all work packages, and many of the preliminary items, should be let following competitive tenders.
- *Inflated costs* – There is a tendency for contractors/managers to wish to obtain tenders only from established sub-contractors with a reliable record. This is good for ensuring satisfactory performance but if cheapness is the most important consideration to the client, regardless of standard of workmanship and finishing times, the lowest cost is unlikely to be achieved.

- *Lack of contract sum* – Whilst the estimate of prime cost is a carefully prepared and agreed budget, which is constantly refined and in which confidence will grow as packages are let, it is not a commitment and some clients are unhappy about entering into a contract which does not have a contract sum or finite commitment.
- *Increased risk* – In order to secure the contractor's/manager's allegiance it is usual that he carries very little contractual risk, that is, greater risk is carried by the client. For example, in the event of a default by a sub-contractor, the client will usually be responsible for all of those effects on the cost and time of the project which cannot be reclaimed from the sub-contractor.
- *Duplication* – There is a possibility of duplication of some preliminary items in cases where sub-contractors are each made responsible for an item which would otherwise have been provided by the main contractor. Scaffolding is an example.

Construction management

As noted earlier in this chapter, there is only one significant difference between management contracting and construction management – that, in the case of construction management, the parties to the package contracts being the employer (instead of the management contractor) and the package contractor. This was illustrated in Figure 12.1. None of the package contracts are entered into by the construction manager – all contracts are with the employer. This leads to some additional advantages, among which are the following:

- *Increased commitment* – Whilst the construction manager employed will very likely be a company or partnership, it is the individual manager responsible who will affect the success of the project. He, not being a party to the package contracts, has virtually zero risk and therefore his sole allegiance is to the client and to the project.
- *Failure of construction manager* – If, despite all the interviewing and checking, the construction manager fails to perform, a replacement can be put in place with comparative ease – certainly more easily than replacing a management contractor party to numerous package contracts.

These additional advantages have led to many employers preferring construction management contracts over management contracts. There are, however, also some additional disadvantages to the use of construction management, among which are the following:

- *No sanctions* – Since there is not usually a liquidated damages provision within the construction manager's contract of engagement, depending on the precise terms of engagement, the client may have no real sanction in the event of non-performance of the construction manager, short of replacing him.
- *Additional involvement* – Since the employer is a party to a large number of direct contracts he is responsible, with the construction manager's assistance, for the risks associated with this and the resolution of any disputes. For this

reason this method of procurement is probably only appropriate for clients who regularly commission building work, have a measure of in-house expertise and wish to be involved in the detailed day-to-day progress of the works. However, since the employer should have a full complement of professional advisers in his team, this should not be a problem.

Use

Management or construction management contracts will very likely be appropriate in the following situations:

- where it is necessary to start work on site before the design is fully developed, usually because speed is the main priority;
- where employers need maximum flexibility to make changes to the building or to engage their own direct contractors during the course of construction;
- where timely completion of the project is of a higher priority than cost; and/or
- on large and complex projects where the enhanced team effort pays dividends and the extra cost is comparatively insignificant in relation to the total project cost – on small projects these contracts can be very expensive.

Programme

As noted above, management and construction management contracts are often used when timely completion is more important than the lowest price.

Single-stage competitive tendering

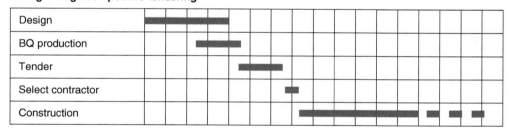

Management contracting/construction management

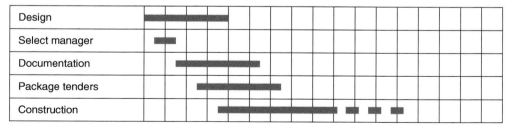

Figure 12.2 Comparative sequence of events.

Figure 12.2 shows a diagrammatic representation of the different sequence of events between single-stage selective tendering contracts and management and construction management contracts. The vertical divisions of the matrix are not intended to indicate any particular time interval – they are provided purely to assist visual comparison of the alternatives. Neither is the length of the activity bars necessarily significant, except in allowing a comparison to be drawn between the two systems. In practice, the duration of any activity must be agreed upon between the participants so as to achieve the employer's expectations. Nevertheless, Figure 12.2 indicates that a considerable time saving can be achieved using a management or construction management contract – although this may not be the main driving force behind the selection of this method of procurement.

Chapter 13
Design and Build Contracts

A design and build (DB) contract is a contractual arrangement in which the contractor undertakes both to design and to construct a project for a single contract sum. At the extreme, DB projects can require that the contractor purchase land, obtain planning permissions and consents, finance, design, procure and construct. These contracts are known as 'turnkey' contracts and derive their name from an employer wanting to have little more involvement than simply turning the key to begin the use of a completed project. Large infrastructure projects such as power stations are often procured this way.

DB has emerged to be the most frequently used procurement method today. The 2010 industry survey for the Royal Institution of Chartered Surveyors established that approximately 33% of the total value of the projects undertaken was procured this way.

Contractual arrangements can vary considerably. For example, during the 1970s and 1980s when DB was in its infancy, the contractor would have undertaken the complete design and construction, from inception to completion, using in-house designers and construction staff. More recently, employers have instructed independent design teams to take the brief, prepare preliminary designs, cost the project and prepare fully detailed designs and employer's requirements before entering into a contractual arrangement whereby the contractor assumes full responsibility for the complete design and construction of the works. At the point when the design is handed over to the contractor it can be anywhere between 0% and 100% complete, such is the flexibility of DB. In some instances, the independent design consultants employed by the employer may switch allegiances at a convenient time (in most cases when the design is complete) to be employed by the contractor. This contract arrangement is called novation.

DB contracts can be negotiated but are predominantly subject to competitive tendering as employers usually see a fixed price commitment from the contractor as one of the main advantages of the system.

The Aqua Group Guide to Procurement, Tendering and Contract Administration, Second Edition.
Edited by Mark Hackett and Gary Statham.
© 2016 John Wiley & Sons, Ltd. Published 2016 by John Wiley & Sons, Ltd.

The contract

There are a number of standard forms of construction contracts which cater for DB procurement but the most widely used is that of the Joint Contract Tribunal (JCT), namely the DB Contract. However, the JCT also produces the Major Project Construction Contract and the Constructing Excellence Contract which may also be used for DB procurement. The JCT's main traditional construction contract, the Standard Building Contract (SBC), also provides for an element of contractor design under a contractor's designed portion (CDP) option contained within the body of the SBC form. The CDP essentially incorporates the design provisions of the DB contract.

The 2011 suite of contracts also introduced standalone contractor design forms in respect of the Intermediate and Minor Works contracts (ICD and MWD, respectively).

The DB contract embodies some very different concepts to those of the traditional JCT contracts for use with consultant design. The key agreements are set out in the first three recitals in the Articles of Agreement, as follows:

- First recital. This identifies the works, their location and the fact that the employer has supplied to the contractor documents identifying his requirements. These are referred to as 'the Employer's Requirements'.
- Second recital. This confirms that the contractor has submitted proposals for carrying out the design and construction of the project referred to in the first recital. These are referred to as 'the Contractor's Proposals'. The recital states that they include a statement of the sum required for carrying out the works and an analysis of that sum. This is referred to as 'the Contract Sum Analysis'.
- Third recital. This records that the employer has examined the Contractor's Proposals and Contract Sum Analysis and is satisfied that they appear to meet the Employer's Requirements.

The contractor's obligation is to design and complete the works in accordance with the details contained in the Employer's Requirements and the Contractor's Proposals. The Contractor's Proposals should expand upon the details contained in the Employer's Requirements, which can vary from a simple schedule of accommodation requirements through to a fully designed scheme.

It is important to note that, if there is a discrepancy between the Employer's Requirements and the Contractor's Proposals, the contract dictates that the Contractor's Proposals take precedence. The contract also covers discrepancies within the individual documents – these are usually resolved in favour of the 'innocent' party.

Under the contract, the contractor is responsible for the design and bears the same professional liability as a consultant designer. The contractor is therefore bound to exercise the reasonable skill and care expected of a competent designer. The contractor is also responsible for full compliance with statutory requirements.

The contract does not provide for the employer to appoint an architect or quantity surveyor as does traditional procurement. There is instead an Employer's Agent who acts on behalf of the employer and receives or issues applications, consents, instructions, notices, requests or statements in accordance with the conditions.

The contract includes two options for dealing with payments to the contractor and the parties should select the preferred method, and provide the appropriate information, before signing the contract. The options are:

- Stage payments (Alternative A), on application of the contractor in accordance with the schedule of payments included in the contract particulars.
- Periodic payments (Alternative B), on application by the contractor in accordance with the period set out in the contract particulars.

The stages, under Alternative A, might include an initial payment upon receipt of planning consent (if that has been the contractor's responsibility) or commencement of work on site and then at defined stages of the project such as completion of the foundations, frame and so on. Under Alternative B, the first periodic payment would have to include the contractor's costs in the early stages of the process and then continue at intervals not exceeding one month based upon valuations of the work completed.

Where to use DB (and when not to do so)

The manner in which DB contracts are established can profoundly affect the quality of the final product. The recommendation to use the system must therefore be based not only on the type of building required but also upon the client's expectations with regard to programme, cost in construction, cost in use, level of specification and quality of design.

DB contracts are certainly useful for:

- Standard building types, especially for industrial or warehousing use, and particularly where early return on capital investment outweighs considerations of design excellence.
- Buildings using proprietary systems where the manufacturer of the system might well become the main contractor, for example, as applied to repetitive housing or low cost hotels based on the assembly of factory made pods.
- Building types in which some contractors have become specialists – for example, highly serviced health care or laboratory buildings in which complexity justifies placing the contract with the organisation with specific expertise of constructing such buildings.

Frequently, systems of construction apply to the basic structure only, so that the choice of layout, finishes and external works can be designed to suit the client's needs without financial penalty. There may well be economic advantages to the client where a contractor's proprietary system can be used without compromising the brief.

In the public sector, where the Private Finance Initiative remains a means of procuring a building or project, bids are usually submitted by consortia, often led by the contractor, as the member of the team most likely to have the resources to fund the extensive pre-contract and tender work. Design, build, finance and operate schemes work in a similar way. The risk involved in speculative work on these projects is considerable and can affect all members of the project team – although the contractor usually takes the largest share, and receives the greatest part of the reward if the bid is successful. In such circumstances, the contractor will expect to enter into a DB contract with a corporate body especially established to procure the project. The question then arises as to the nature of the relationship between the design team and the contractor, a subject considered below.

There are, however, other types of project for which DB is a less suitable form of procurement:

- where architectural quality is of overriding importance (the client may even wish to instigate an architectural competition);
- where clients require a building tailored to their special requirements;
- where there are complex planning or environmental issues;
- where complex refurbishment work (particularly of historic buildings) is required, as frequent or unexpected variations often arise; and/or
- where the brief cannot be defined, or the building function is of great complexity, so that a protracted period of research and investigation is necessary at the outset and might even continue once work has commenced.

Whilst the issues above might be construed as a disadvantage to DB as a method of procurement, there are nevertheless means of resolving most of them. An employer may use a design team in the traditional sense to develop the brief before handing a completed design over to the contractor through the mechanism of novation. The use of DB contracts is principally a risk-driven process where the contractor assumes the legal risk for both design and construction.

Managing the design process

The key to any DB contract is the brief or 'Employer's Requirements'. This may be in the form of an outline specification, a performance specification, drawings in varying degrees of detail or drawings and an outline specification in combination.

The employer often appoints a design team to assist in compiling this brief and preparing preliminary designs and costings before seeking a contractor by competition or negotiation to proceed on a DB basis. The costing element of this exercise is important as it is vital that the contractors tender on a basis that the client knows is within the budget.

DB is an expensive procurement route for contractors to tender in competition, as each has to work out a detailed design to suit the brief and a price to go with it. Clearly this can be an uneconomic use of resources, which will reflect on the ultimate cost to the employer – especially if too much detail is required as part of the competition process. Before contract, therefore, contractors should go no further than an outline scheme design based upon a specific and concise brief. Detailed design and a fully detailed specification come later.

Once the contractor has been appointed and has assumed responsibility for the detailed design, the decision must be made as to the future role of the employer's consultants. When the contractor uses an in-house design team to develop the details for construction purposes, the original designers and quantity surveyor might be retained by the employer to monitor standards and supervise payments. Alternatively, the original designers might be 'novated' to the contractor, leaving the quantity surveyor to give cost advice to the client and/or act as the Employer's Agent.

Novation

Novation is a legal term used to describe a procedure whereby one contract is substituted for another – commonly where a contract between two parties, A and B, is replaced by a contract between parties A and C. Novation is widely used on DB contracts, when the contract between a consultant and the employer is replaced by a contract between the consultant and the contractor. However, as noted below, the terms and conditions of the substituted contract are, of necessity, different to some extent.

Novation will normally take place as the contract is let. The Employer's Requirements, issued as part of the tender documents, should clearly state that the consultant's appointment is to be novated, and include the precise terms and conditions of the consultant's original appointment. Since the consultant's original appointment will frequently include the provision of services which have been completed, or which are no longer relevant to the new appointment, the terms and conditions of the new contract are bound to be different. For example, feasibility studies and sketch proposals, forming part of the original appointment, will already have been completed and will not be required as part of the new contract and the contractor will not necessarily require a post-contract management service. A schedule identifying the services which are to be provided under the new appointment should be drawn up and agreed by the consultant, employer and contractor.

Difficulties often arise when employers fail to realise that novated consultants no longer have duties to perform for them. In practical terms, the consultants are no longer employed by the employer and are therefore no longer able to represent the employer's best interests, monitor the quality of construction or deal with payment. These duties must be left to others, such as the Employer's Agent. With due advance planning, however, most potential issues can be anticipated. Depending on the extent of the design responsibility included in the contract and the contractor's own professional indemnity insurance cover, the consultants may be required to enter into collateral warranty or third party rights agreements protecting the employer against damages arising out of design-based failures. These indemnities might only be called upon in the event that the contractor becomes insolvent since, following novation, the employer has a direct contractual link with the contractor for guaranteeing adequacy of the design.

Part II

Evaluation of submissions

When DB is used in its pure form, whereby the contractor undertakes to carry out all of the design, once the DB tenders are returned the evaluation of the contactor's proposals is a complex task. Each contractor will almost certainly have interpreted the brief in a different way, making direct comparison difficult. However, in most cases, the submissions can be compared on a common or equivalent basis by adjusting those items which are provided for in one submission but left out of others and comparing overall value for money. In some cases, this is very simple, for example, different floor coverings can easily be evaluated or costed and therefore compared and contrasted.

Design qualities may be a significant factor and in the case of a design competition are obviously paramount. In this scenario, evaluation is very subjective and cannot be expressed in purely monetary terms. Sound professional advice is necessary from independent architects, engineers and quantity surveyors.

After selection there will be a period during which the contractor will be refining the details of the design, preparing documentation for obtaining sub-contract tenders and mobilising site resources. Frequently, the period of development of the design details will bring to light matters which can be incorporated in the building without penalty and for the benefit of the employer. All these matters must be taken into account when finalising the formal contract documents in support of the agreed contract price.

Post-contract administration

Although the total responsibility for carrying out the design and construction work rests with the contractor, the employer is normally well advised to monitor the contractor's performance to ensure that the specification is adhered to, as set out in the contract documents, and that the workmanship is to the requisite standard. If the original consultants have not had their contracts novated to the contractor they will be able to do this; if novation has taken place the employer must rely on others.

Although there is no provision in the contract for this role (in terms of an architect or a clerk of works), the Employer's Agent has reasonable access to the works and the employer has powers to order the opening of covered work for testing.

Financial administration

Care must be taken to ensure that the contractor is paid the correct amount for interim and final payments. Again the employer will be well advised to monitor the contractor's performance, for under this contract all the financial calculations are undertaken by the contractor. The contract does not make provision for an employer's quantity surveyor, although the employer still has the right to challenge the contractor's calculations.

The contractor is charged with making applications for payment, using either the stage payment or periodic payment system. Valuations made by the contractor will include the valuation of variations which must be priced according to the valuation rules laid down in the contract. This, as with other JCT contracts, is a unilateral activity but in this instance, of course, it is the contractor who does the calculations and not the employer's quantity surveyor.

It is usual for the employer to appoint a quantity surveyor as financial administrator, and preferably to retain the services of the quantity surveyor who has been involved in the preparation of the brief, or Employer's Requirements. It is not uncommon for this appointment to coincide with that of the Employer's Agent.

Programme

The main driving force behind the use of DB contracts is probably the desire to have a single point of responsibility, closely followed by the hope of a reduced risk of costs rising. However, a reduced timescale for the project is also a possibility.

Figure 13.1 shows a diagrammatic representation of the different sequence of events between single-stage competitive tendering and DB contracts. The vertical divisions of the matrix are not intended to indicate any particular time interval – they are provided purely to assist visual comparison of the alternatives. Neither is the length of the activity bars necessarily significant, except in allowing a comparison to be drawn between the two systems. In practice the duration of any activity must be agreed between the participants so as to achieve the employer's expectations.

In Figure 13.1 the overall project timescale is shown as the same for both systems but, as noted above, this is not intended to be significant. Depending on the

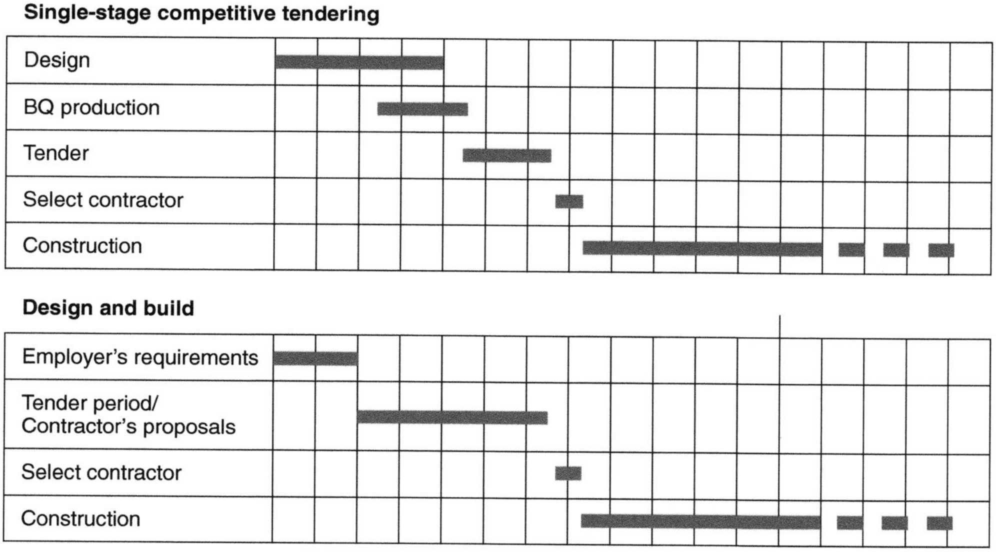

Figure 13.1 Differing processes compared.

size and complexity of the project, and the client's priorities, the overall timescales can be varied. However, many DB specialists claim that programmes can be shortened because the process is closely co-ordinated within their single responsibilities and because of the shortened learning curve – this is because team members are used to working together.

Advantages and disadvantages

Advantages

- Early certainty of overall contract price is obtained, as long as the JCT form of contract is used.
- Responsibility for the design, construction and the required performance of the building lies entirely with one party, the contractor.
- DB imposes a discipline, not least on the employer, to define the brief fully at an early stage. Given this, the advantages of overlapping design with construction can lead to a shorter project duration.
- Given the higher degree of co-ordination at an early stage, through the single point of responsibility, variations during construction tend to be fewer and the risk of post-contract price escalation is thereby reduced.

Disadvantages

- DB, by its very nature, is a rigid system which does not allow the employer the benefit of developing requirements and ideas, despite the facility for accommodating variations.
- Where several tenders are invited, comparison can be difficult as the end product in each case is different, and the final choice will be influenced by subjective judgment.
- Any variations required by the employer after signing the contract can prove expensive and difficult to evaluate.
- There is always a risk with regard to the quality of work. If the original brief is not precise and the specification offered by the contractor is equally vague, there is a temptation for the contractor to reduce standards.
- If the design team's contract is novated to the contractor it is possible that a conflict of interest may arise.
- Following novation of the design team's contract to the contractor, employers have no direct input for checking or improving quality unless they appoint other consultants.
- Most DB contracts are qualified in some respects (e.g. by ground conditions or the inclusion of provisional sums) which to a certain extent negates the employer's ideal of an early known financial commitment.

Chapter 14
Continuity Contracts

In Chapter 4, several aspects of the economical use of resources were considered. Among these was the subject of continuity. Continuity has two characteristics which are of interest in this context: first, economies of scale and, secondly, the effect on the learning curve. Both characteristics have potential favourable cost implications – the first directly in reduced material costs derived from greater buying power, and the second indirectly through a reduction in timescales and thereby reduced labour costs.

In previous chapters, various methods of procurement and contractual arrangements have been discussed in relation to single projects or construction sites. In circumstances where continuity of work is possible, advantage can be gained by adopting a different approach – by linking projects or work packages together.

The greatest scope for applying such concepts is obviously with clients or agencies that have a regular or a continuous building programme. This can be for one or more new buildings and/or in maintenance contracts which require the same repetitive type of work to be undertaken on a number of buildings and/or for a certain period of time.

In this chapter, the purpose and use of three types of continuity arrangements are compared and their operation is described in order to illustrate how benefits can be gained. The three types are:

- serial contracting;
- continuation contracts; and
- term contracts.

Serial contracting may be incorporated into a framework/partnering approach. The JCT suggests that

The Aqua Group Guide to Procurement, Tendering and Contract Administration, Second Edition.
Edited by Mark Hackett and Gary Statham.
© 2016 John Wiley & Sons, Ltd. Published 2016 by John Wiley & Sons, Ltd.

Framework agreements enable project participants to take a longer term view – to build and develop relationships – to focus on the client's needs – to learn from experience – to invest in product development and more effective and efficient processes – to invest in people – and as a consequence of all of the foregoing, generally enhance the reputations and commercial opportunities of all concerned.

The aim of framework agreements is, therefore, to promote and aid the use of collaborative working.

Serial contracting

In many programmes of building work, such as providing several supermarkets for a supermarket chain or refurbishing a chain of high street shops for a retailer, there is an element of continuity. The object of any tendering procedure should be to ensure that building resources, including associated professional resources, are used as economically as possible. Where there is continuity these resources may be used more economically and efficiently if all the work is carried out by the same contractor, rather than by having a different one for each site or operation. It is important, therefore, to identify where the continuity lies and to evaluate the benefits that will accrue if one contractor undertakes the work instead of several contractors.

Through economies of scale driven by the repetitive nature of the work, contractors with the knowledge of security of future work may work for reduced margins since wider industry market risk is reduced. In much the same way as a client benefits financially through continuity, the contractor can also secure financial benefits through greater efficiency in the control of the supply chain.

However, it is not enough merely to establish that there is a saving in resources. As the clients provide the continuity by virtue of the programme, they should gain more than just financially by securing consistent quality and good working relationships. It will be appreciated that in any situation where one contractor is to do a series of jobs for a client, a good relationship becomes particularly important.

Serial contracting has been broadly defined as an arrangement whereby a series of contracts is let to a single contractor, but further definition is needed to distinguish it from other types of continuity arrangement. In a serial contract the approximate number and size of projects is known when the interest is canvassed and offers are obtained. The serial tender is a standing offer to carry out a series of projects on the basis of pricing information contained in competitive tendering documents. The use of contracts under a framework agreement is flexible. For instance, the JCT Framework Agreement was designed such that it could be let in conjunction with any of the standard building contract, design and build, intermediate building contract, intermediate building contract with contractor's design, minor works (MW) and minor works contract with contractor's design. The pricing mechanisms under serial contracts may consist of master bills of quantities, schedules of rates and/or employer's requirements, specifically designed to cover the likely items in the particular projects envisaged.

The series will usually consist of a minimum of three projects; the number will normally be known at the time of the tender, although further projects may be added by agreement at a later stage. Final designs will not necessarily be ready for all of them at the tendering stage but clients and their architects will have a good idea of the typical requirements and these will be reflected in the tendering documents prepared by the quantity surveyor.

Exact quantification will be achieved as the design for each project in the series is finalised. Normally a separate contract based on the original standing offer will then be negotiated and agreed for each project, based on standard items extracted from the master documents. All new or 'rogue' items will be negotiated for each contract as they occur.

Purpose and use

Serial contracts are ideally suited to a programme of work where the approximate number and size of projects to be contracted is known at the time of going to tender. For example, in refurbishing a series of high street shops for one client it will normally be the quantity of work which varies rather than the quality or specification. Similarly, although supermarkets for a chain retailer may vary in size, in a new building programme even the largest is likely to be less than three times the size of the smallest. Moreover, the sizes of most of the projects are likely to be far closer to each other than this. Therefore the same type of contractor management is likely to be needed for them all. It is probably this latter point which should be regarded as one of the governing criteria.

Consideration should be given to the choice of this method of contracting in times of high inflation, or when there is a building boom, when it may be difficult to secure a firm price commitment over a lengthy period of time. Equally, one might not want to get locked into a situation whereby one is paying high rates in an economic downturn, so a balance must be struck with reasonable expectations.

Operation

On serial contracts the tendering procedure will normally be on the basis of a standard set of documents which may comprise most of the standard items envisaged in the various projects in the series. Individual projects will be controlled by having separate contracts negotiated on the basis of a standard schedule of rates, after which work will proceed very much as for a normal contract.

It is desirable to take great care in the selection of the contractors for the tender list before formal tendering takes place. A mistake made when dealing with a series is likely to have more serious results than in the case of a one-off contract, so it is much more important to establish the financial and physical resources of the contractor. It would be advantageous to make provision within a framework agreement for the client to be able to withdraw from a contract if performance is found to be below par. Such performance can be measured against key performance indicators set before a framework is formally entered into. A further point to note is that small errors of documentation or detail, which may not amount to much on one contract, may be multiplied under a framework.

Comparing the contractor's capacity with the likely programme of work also requires great care. Although the total series may be large, the individual contracts may be relatively small. It is possible, therefore, that a contractor who might not be able to cope with such a large contract on one site would be able to cope with a number of smaller projects spread over several sites and, possibly, over a longer time. It is important, however, that the contractor's physical resources should not be stretched to the point where it is difficult for all the work to be completed. Any unprogrammed overlaps in projects could present a contractor with resourcing problems. If significant changes in the overall programme occur, it may be worth considering whether or not to retender.

It is likely that there will be a stage before formal tendering, where interviews are held with possible tenderers, references sought and a short-list is drawn up on the basis of the information obtained.

The result of competition should be that the saving in building resources arising from the continuity is passed to the client. It is emphasised, however, that this is only possible where the extent of the work to be covered by the serial contract is known sufficiently for reasonable estimates of prices to be given in the original standing offer. It should be noted that the savings derived from serial contracting are evidenced in all projects – the savings are contained in the rates in the master bill used to price all individual contracts.

Continuation contracts

A continuation contract differs from a serial contract in that it results from an ad hoc arrangement to take advantage of an existing situation. Thus, if a contract is already going ahead on a particular phase of a business park site, the opportunity may arise at that point to commence a further phase with similar requirements. In this case there may have been no standing offer to do more work, the original tendering documents having been conceived only for one particular phase. However, those documents can provide a good basis for a continuation contract. Alternatively, it is possible to make provision for continuation contracts in the tendering documents for the original project, and this is sometimes done. There is, however, no contractual commitment, and in the event a continuation contract may not arise.

In either case the contractor has to price the tender documents for the original contract on the basis of getting that particular contract only. The continuation contract, if and when it arises, is then dealt with separately – and it is only at this stage that savings arise.

Purpose and use

As with serial contracts, continuation contracts are best suited to situations where the scope, style and nature of the work are very similar. The more variations there are between the projects, the more variations there will be in the contract documentation. This will result in an increased number of purely negotiated costs (rather than adjustments of competitive prices) and could also lead to organisational problems on site.

The purpose of a continuation contract should be to seek a financial advantage for the employer. Continuation contracts offer the employer the following potential benefits:

- a faster start on site, resulting from the shorter pre-contract period;
- a competitive basis for pricing, resulting from the initial tender;
- an experienced contractor, who has identified (and solved) the construction problems of the previous contract (this represents perhaps both a cost and a time benefit); and
- re-use of the contractor's site organisation and management team.

Whilst circumstances do arise where the use of continuation contracts might be appropriate, such as phases of a business park, the practicalities of a situation often preclude this. For example, whilst the style of the work is very similar, the projects would not necessarily be of a comparable size. Additionally, uncertainties such as planning approvals, existing leases/tenancies and enabling/preparatory works often make site acquisition and site handover ready for construction a precarious job and these factors create obstacles to continuity.

However, provided that there is a degree of parity between the projects, a continuation contract may be appropriate in circumstances where serial contracting is not.

Operation

In continuation contracts, the tendering documents will obviously be closely related to the documentation for the original project on which the continuation is to be based. The contract sum is negotiated on the basis of the original contract but with two substantial adjustments.

First, the contractor may require increases in the cost of labour and materials to be taken into account, to allow for the time difference between the two tenders. The only exception to this might be where the original contract was let on a fluctuations basis – the continuation contract could then, if necessary, be from that same base and the increases would be covered entirely by the fluctuations clause. The same would apply, of course, to any decreases.

Secondly, employers will wish the benefits of the continuity they have provided to be passed to them in the continuation contract. These will be of two kinds:

- Productivity in the construction industry generally has increased year by year and, therefore, if the continuation contract is timed, say, nine months to a year later than the original contract, there should be a saving resulting from increased productivity.
- There is a further saving arising from the economies inherent in the same contractor doing similar work for the same architect and client. This element varies from one contract to another, but may be quite substantial where similar houses or flats are being built on the second site.

Part II

The matter of measurement of productivity is difficult, and requires much research. A lot will depend on the information available to the quantity surveyor, and that person's skill in analysing and assessing elements which affect productivity and in negotiating with the contractor on this basis.

One of the prerequisites for the successful negotiation of a continuation contract is parity of information between client and contractor. This involves the contractor disclosing the details of how the tender for the original contract was built up. If the contractor is not prepared to do this, it may not be prudent to proceed on this basis.

Once the continuation contract has been negotiated, a figure agreed and a contract signed, the project will proceed as any other. There is one factor, however, that should be considered, particularly where it is envisaged that several continuation contracts may arise from the original. The contractor should be given an incentive to make cost reductions. Where continuation contracts are being considered, the contractor might well be given the full benefit of any reduction in cost he makes on the first occasion, even though this involves a change in specification (but not, of course in performance), with the proviso that the full amount of the cost reduction be allowed to the client in the succeeding continuation contract. This situation arises particularly where the contractor is involved in a considerable amount of design work, and where the detailed specification is very much related to his production requirements.

Term contracts

Term contracts differ from both serial and continuation contracts in that they envisage a contractor doing certain work for a period of time or term. In this situation, the contractor agrees a contract to do all work that he is asked to do within a certain framework and during a given period. Term contracts are generally set to run for a period of 12 months, but a longer period is likely to result in better tenders. Experience suggests that a term of up to three years can generally be expected to promote confidence within the period of the contract. However, the volatile nature of building costs is such that provision should be made for the employer to test the market at shorter intervals. Thus a period of two years may, on balance, be regarded as the optimum for such contracts.

Purpose and use

Term contracts are well suited for use with the management of large and continuing programmes of day-to-day reactive building maintenance. The management and control of such programmes can be an extremely complex task and, in view of their potential for flexibility, term contracts are perhaps best suited to the problem.

Care should be taken both in the compilation of the tender documents and in their evaluation. Generally speaking, the main tendering document will be a schedule of rates. Additionally, it can be helpful to prepare a bill of provisional quantities, based on the previous years' workflow, to give an indication of the likely volume of work anticipated on specific trades over the term of the contract.

As always, it is advantageous if contractors of like characteristics and performance can be selected to submit tenders, and it will usually be an advantage if multi-trade contractors can be utilised in view of the many operational, as opposed to single-trade, jobs to be found in building maintenance.

Term contracts may also be used for painting and redecoration, roadworks and other specialist trades.

JCT Measured Term Contract

At the request of its constituent bodies, the JCT produced its first Standard Form of Measured Term Contract in 1989. It is specifically intended for employers who have programmes of regular maintenance and MW, including improvements.

This form requires the employer to list the properties in the contract area and the type of work to be covered by the contract, as well as the term for which it is to run. The employer is required to estimate the total value of the contract and to state the maximum and minimum value of any one order. There is provision for priority coding of orders to deal with emergencies and specific programming requirements.

Payment is made on the basis of the National Schedule of Rates, or some alternative priced schedule, and the contractor quotes a single percentage adjustment to the base document, or perhaps different percentages for different trades. Provisions are included for fluctuations and dayworks. Measurement and valuation of individual orders can be carried out by the contract administrator or by the contractor, or can be allocated to either party according to value.

It is also possible to compile an ad hoc schedule, relative to any employer's particular range of work.

Operation

Term contracts are ideal for carrying out maintenance and repair work. In this type of work, the individual project can be very small, say from £500, while the largest may be in the region of £30,000 or more. Tenders can be sought using a large number of methods but two of the commonest are as follows:

- the employer can determine the unit or operational rate for each item and the tenderers then offer to do the work on a plus, minus or 0% basis to reflect their bid; or
- tenders can be sought on the basis of blank schedules which the tenderers then price.

Analysis of the tenders will usually be found to be easier if the first method is used; depending on the number of tenders received, the analysis of priced schedules can be a daunting task. Experience will suggest the best method for individual circumstances.

Additionally, as part of the contract, tenderers may be required to meet specified response times for various categories of work, including emergency items. The ability to meet these response times will be an important consideration in the selection of the successful contractor(s).

Contracts can be entered into with one or more contractors and the work placed on the basis of cost, committed workload and performance. Generally speaking, the more contractors available to do the work, the smoother the operation of the maintenance programme. Orders for work are issued from time to time; the work is agreed and the contractor paid accordingly. The number of orders issued annually under a maintenance term contract can run into many thousands. To ensure the workload can be dealt with smoothly and expeditiously, continual monitoring of the contractors' performance will be necessary and orders to the various contractors regulated accordingly.

Orders should be given a priority rating which will require contractors to attend and carry out the work within a specified time. An incentive scheme will be found to be helpful to achieve this. Where the work is completed within a specified time, an attendance payment can be made. Alternatively, and perhaps more effectively, where the contractor fails to complete the job on time, the number of orders can be reduced until performance improves again. Additionally, a plus payment can be made for attendance to emergency call-outs. Emergencies which relate to health and safety matters can arise at any time and will usually require quick attendance – initially to make safe only, the permanent repair being dealt with separately.

Receiving repairs requests, deciding on remedial action, determining priorities, allocating work, monitoring attendance and completion times and inspecting and certifying payment can be a complex exercise. The use of a sophisticated computer programme, which should be based on a property register and linked with a comprehensive schedule of rates, together with an order generator and accounts for payment, will be found to be invaluable in the smooth running of a maintenance-based term contract.

For an employer with a large estate and a large number of MW of this nature to be carried out, a term contract offers a method of controlling the work with a measure of accountability and a simple method of procurement for the individual projects.

Chapter 15
Partnering

Partnering is perhaps seen as being a relatively new approach to the procurement process, in which an attempt is made to improve relationships and performance for the benefit of clients and the other members of the team – what is termed, in management parlance, a 'win-win' situation. Many clients are dissatisfied with the results of individual competitively tendered projects. These contracts have often underperformed in terms of time, cost and quality and have led to adversarial attitudes from all parties.

Since reference to partnering was made in Sir Michael Latham's report *Constructing the Team* and Sir John Egan's report *Rethinking Construction* there has been a steady line of publications advocating its use. These, however, are only manifestations of a process which has been developing for some considerable time, although the name 'partnering' may only have been applied to it recently. It is said that the oil, chemical and power generation industries in America first used the system, while in the UK the British Airports Authority's framework agreements and the National Health Service's Procure 21+ can be considered as forerunners to what is now recognised as partnering.

All these reports have thrown out the challenge to the industry – to improve relationships and to deliver measurable improvements in time, quality and cost.

A definition

The terminology encountered in this field: 'project partnering', 'strategic partnering' and 'first, second or even third generation partnering', or even 'collaborative working', gives a clue as to the variety of interpretations placed on the term 'partnering' and therefore the impossibility of a finite definition.

The Aqua Group Guide to Procurement, Tendering and Contract Administration, Second Edition.
Edited by Mark Hackett and Gary Statham.
© 2016 John Wiley & Sons, Ltd. Published 2016 by John Wiley & Sons, Ltd.

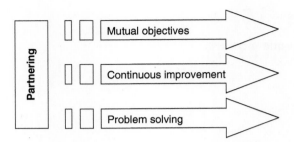

Figure 15.1   The three fundamental characteristics of partnering.

Partnering has been described as a structured management approach which facilitates teamworking across contractual boundaries by integrating the project team and smoothing the supply chain. The three fundamental characteristics of partnering (see Figure 15.1) are:

- formalised mutual objectives (which may be binding or non-binding) of improved performance and reduced cost;
- the active search for continuous measurable improvement, which is perhaps measured against industry key performance indicators (KPIs); and
- an agreed common approach to problem resolution.

Partnering should not be confused with other good project management practice, or with longstanding relationships, serial working, negotiated contracts or preferred supplier arrangements, any of which may be present but all of which lack the structure and objective measures that support true partnering.

The essential characteristic of partnering is the commitment of all partners at all levels to make the project a success. The result is that the partnering agreement between the parties drives the relationship, rather than the contract documents. This is not to say that the contractual relationships are not important – they must be in place for the day-to-day running of the project. Standard forms of contract can be used in partnering, although it is as well to check the specific terms of any document to make sure that they do not cut across the aims of the partnering agreement.

Partnering can be applied to single jobs, often then termed 'project partnering', but the benefits of partnering tend to be cumulative, so that significantly greater benefits arise over several projects – a situation often termed 'strategic partnering'. It is also considered that the benefits to the client are further enhanced if partnering is applied throughout the supply chain, thus covering the employer, the professional advisers, the contractor and sub-contractors and the suppliers. If the client's workload is sufficiently large or geographically dispersed, there is no reason why several different partnering agreements should not be set up, running concurrently, with different groups of partners.

It should be borne in mind that, according to EU procurement regulations for public sector procurement, partnering can only be applied to projects after the award of the contract. The European Construction Institute's document *Partnering in the Public Sector: A Toolkit for the Implementation of Post-Award, Project Specific, Partnering in Construction Projects* is particularly relevant in this context.

The key to partnering is a change in mental attitude, a new approach based on honesty, co-operation and deeds (not just words). Without a fundamental change in attitude incorporating these three ingredients, no partnering agreement will work, or achieve the claimed significant improvements in performance.

When to adopt a partnering approach

Partnering is applicable to all projects in terms of supply chain management and being able to obtain a mutual benefit for both parties. However, there is perhaps more scope for partnering in circumstances where there is a high likelihood of inefficiency otherwise undermining the project. Partnering is said to be particularly appropriate in situations where:

- the project is technically complex and difficult to specify;
- the business requirements require advice upstream and downstream of the supply chain;
- the client has projects of a similar nature, giving scope for continuous improvements in terms of time, cost and quality; and
- construction conditions are uncertain and solutions are difficult to foresee, giving rise to an element of joint problem solving.

The agreement

A partnering agreement may take the form of a verbal agreement, a broad written arrangement on a single piece of paper or a more detailed and formal document. Each situation will have different aims and requirements, which will dictate the appropriate method of setting up the partnering agreement. Care has to be taken, however, to ensure that a detailed written agreement does not lead to an adversarial attitude if things go wrong. One might also question what benefit there is in a document which comprehensively codifies the expected behaviours of the project participants because the need to do so may underscore the point that the behaviours are not those which would instinctively be followed.

The intention of a partnering agreement is to establish a framework, based on mutual objectives and agreed procedures, allowing the participants to maximise specific business goals. A partnering agreement should consider the following areas:

- The formulation of a statement outlining the overall philosophy of the parties – that of working in good faith.
- The setting of realistic and achievable targets, with a procedure to monitor and review these as necessary.
- The agreement of a method where open-book costs can be established and savings can be shared by all participants.
- A procedural framework, including the definition of the parties' roles, responsibilities and lines of communication.

Part II

- An agreed procedure for the resolution of disputes, without damaging the original objectives of the partnering agreement.

JCT Partnering Charter

The JCT publishes a non-binding charter PC/N for situations where parties do not wish to enter into a legally binding agreement but do want to create a collaborative working environment. The charter is a short document recording the team's commitment to work together to produce a project that meets the client's needs, the clients expectations in respect of quality, be within the agreed budget/price and be completed according to the project programme. It requires the signatories to act:

- in good faith;
- in an open and trusting manner;
- in a co-operative way;
- in a way to avoid disputes by adopting a no-blame culture;
- fairly towards each other; and
- valuing the skills and respecting the responsibilities of each other.

JCT Framework Agreement

The JCT framework agreement (FA) is a legally binding 'umbrella' agreement. Framework arrangements are typically used where an employer has a long-term programme of work in mind and wishes to set up a process that will govern separate packages of work that may be required during the period of that agreement. JCT describes the aim of the Framework Agreement in these terms:

> "The main aim of this Framework Agreement is to provide a mechanism for the Tasks to be called off and carried out and also to provide a supplemental and complementary framework of provisions designed to encourage the Parties to work with each other and with all other Project Participants in an open, co-operative and collaborative manner and in a spirit of mutual trust and respect ... "

(Clause 3.1)

There are nine core framework objectives and further provisions advocating mutual benefit for the parties. These are:

- zero health and safety incidents;
- teamworking and consideration of others;
- greater predictability of out-turn cost and programme;
- improvements in quality, productivity and value for money;
- improvements in environmental performance and sustainability, and reductions in environmental impact;

- right first time with zero defects;
- the avoidance of disputes;
- employer satisfaction with product and service; and
- enhancement of the Provider's reputation and commercial opportunities.

Provisions are included for the settlement of disputes which is, perhaps, somewhat of a paradox in terms of the partnering ethos.

JCT constructing excellence

The JCT's Constructing Excellence contracts and guides comprise the following documents:

- Constructing Excellence Contract (CE)
- Constructing Excellence Project Team Agreement (CE/P)
- Constructing Excellence Contract Guide (CE/G)

JCT claims the contract is appropriate where participants wish to engender collaborative and integrative working and for use in partnering. It can be used with lump sum or target cost payment options.

The contract provides for the use of a risk register, risk allocation schedules, and performance indicators.

The team agreement (CE/P) supports the collaborative approach and formalises the integration of the project team. There are optional sections for risk reward sharing arrangements as between the team members.

The partnering workshop

The practical details of the partnering agreement, its procedures and objectives are often worked out in a one- or two-day workshop, where all the participants come together at the earliest possible stage to plan the project. This is the start of the teambuilding exercise and involves the actual personnel to be involved in the project, not just the senior management or marketing staff.

The workshop is often run by professional facilitators and terminates with the drawing up of the project charter, which should be signed by all participants.

The benefits

A properly set up and effective partnering agreement should seek to achieve benefits in respect of time, cost and quality. Such benefits will be derived in a number of ways, based on the following concepts:

- **Reduced learning curve** – Optimum performance and productivity can be achieved much earlier in the process due to familiarity with working practices and mutual confidence in the expertise of other participants.

Part II

- **Reduced abortive tendering costs** – Overheads involved in tendering are reduced. In traditional competitive tendering, a success rate of perhaps one in five or six is not unusual; this adds a heavy burden to overhead costs.
- **Administrative efficiency** – Continuity of personnel and familiar systems of management eliminate teething problems and silly mistakes.
- **Improved communications and decision procedures** – Structured procedures, developed during regular contact, promote confidence and reliability.
- **Improved quality and programming** – An overall strategy giving a proper balance to the client's priorities avoids the achievement of one target at the expense of others, so quality and time are not necessarily sacrificed to cost.
- **Economies of scale** – Advantage can be taken of continuity and preferred supplier arrangements, so partnering exists throughout the supply chain.
- **Continuity of work and personnel** – Long-term relationships can be developed, building confidence and reliability.
- **Risk** – Risk identification is improved, leading to better risk allocation and management.
- **Problem solving** – Problems are identified at an early stage, thus facilitating an early solution and avoiding escalation. If a formal dispute does arise, agreed procedures can be implemented quickly to achieve an early resolution.

The risks

Partnering arrangements are no panacea for the ills of the construction industry and the benefits do not come automatically. Partnering needs commitment at all levels in the organisations involved if the benefits are to be fully realised. However, the parties must be vigilant to prevent complacency setting in. Among the risks attached to partnering, as in many other innovative systems, are the following:

- **Potential lack of accountability** – The absence of competition means that other mechanisms are required to demonstrate the absence of collusion, though quality and audit systems, and the realisation of mutual benefits, should satisfy both parties.
- **Unrealistic targeting** – The search for constant improvement can lead to targets for cost, quality and time being either too low or too high; again the open-book approach to management should provide reassurance.
- **Commercial pressure** – The desire to increase returns on the part of contractors (or their shareholders) may tempt them not to declare openly the true cost savings achieved.
- **Cost attitude** – Having removed, or reduced, the adversarial approach in favour of teamwork, care must be taken to avoid slackness and the taking of shortcuts. As noted above, the ordinary contractual rights and responsibilities must continue to be observed.
- **Change of personnel** – It is impossible to prevent some turnover of staff and there is a danger that new members of the team will not understand the concept, or not be as committed to the project as those they replace.
- **Benefits slow to materialise** – The full effect of the benefits may not appear immediately; indeed some will not materialise unless a number of projects are

undertaken. Careful analysis of the potential benefits, accurate targeting and proper evaluation of the feedback are required to avoid disillusionment, should results be slow to materialise.

- **Programme amendments** – When the employer's requirements or priorities change, or the programme of projects is varied, the mutual goals may no longer be achievable. It is important to analyse fully the effect of any variations in the brief on all members of the partnering arrangement.
- **Dissolution on failure** – The termination of any partnering arrangement, for whatever reason, may lead to financial problems when unravelling complex relationships.

Future of partnering

For partnering to be successful, it requires a more flexible attitude and approach than has traditionally been seen in the construction industry, and an acceptance that organisations and individuals are entitled to receive a reasonable return for their efforts. It also requires people to accept that the cheapest price does not necessarily provide best value for money. A solid base of traditional training and good professional and commercial practice needs to be extended within the industry to allow the more lateral approach of a partnering arrangement to benefit all concerned.

The essence of partnering is in the attitude of the participants, and in their ability to achieve the optimum overall solution, given a particular set of requirements and circumstances, while ensuring that all participants benefit. This will be achieved by matching one's deeds to the words – all too often, people sign up to partnering agreements then act as if they have not, only to express dismay that the partnering agreement has not spared them failure.

Part II

Chapter 16
EU Procurement

Michael Bowsher

Introduction

Procurement in the construction sector in the United Kingdom is to a large extent subject to the rules of EU public procurement law. These derive from a group of EU directives first adopted in the construction sector in 1971, before the United Kingdom joined the then European Economic Community. These directives have subsequently been reworked and expanded on various occasions. The latest reworking of the EU legislation is in three directives adopted in 2014 (and an earlier 2009 directive covering the defence sector). These directives are incorporated into UK law by various regulations.

The regulation of public procurement is currently established in three sets of regulations: the Public Contracts Regulations 2015, and regulations covering the utility and defence sectors. Regulations covering concessions will be forthcoming. Each has been the subject of amendments. Public procurement law in the EU is not, however, a self-standing independent measure. It is part of the wider transnational architecture established by the World Trade Organisation in the context of a plurilateral agreement called the Government Procurement Agreement. The essence of the Government Procurement Agreement is to open up competition in public procurement markets amongst those, mostly industrialised, States that are signatories to that agreement.

In the European Union, while the regulation of public procurement might have been directed most specifically at achieving certain value or policy goals, the genesis and driving force for the existence of the legislation is to drive and support the overall EU goal of achieving and maintaining an internal market.[1] This remains the primary goal of EU, and therefore UK public procurement law.

The Aqua Group Guide to Procurement, Tendering and Contract Administration, Second Edition.
Edited by Mark Hackett and Gary Statham.
© 2016 John Wiley & Sons, Ltd. Published 2016 by John Wiley & Sons, Ltd.

The scope of procurement law

A number of highly complex and technical issues determine whether any particular contract or procurement is covered by the regime. In short, the key questions to be considered are the nature and identity of the purchasing entity, the nature of the contract, the subject matter of the contract, the value of the contract, the geographical scope of the contract and the extent of any cross-border trade in the markets affected by the contract or procurement. All of these matters affect the classification of the contract and determine which regime is applicable to the regulation of the tender process for that contract. There is a further division as already alluded to, between the public sector, the utilities sector and defence sector. Concession contracts, that is, those in which some or all the consideration for the provision of the services or works is afforded to the provider from operating the facility or the service will be subject to a separate new regime following the implementation of the new 2014 directives.

For the purposes of this short chapter, the situation can be simplified on the basis that nearly all public bodies and a large number of utility bodies will be subject to one or other of these regimes in respect of all works contracts valued above £4.3 million or service contracts with a value in excess of £200,000.[2] The legislation as implemented in the United Kingdom is daunting and made even more so when one realises that one has to read that legislation by reference back to the EU legislation, which is even more extensive. The regime would appear to be a highly complex bound regime, and to some extent that is true. However, the core principles dictate the impact of those rules and to some extent at least, a large proportion of the rules are simply either the working through of the general principles or the establishment of a common basis upon which to apply those principles.

The general principles

All contracting authorities must conduct themselves in accordance with the principles of equal treatment, non-discrimination and transparency. EU law probably introduces by way of its case law a further requirement to comply with other EU principles such as the principles of proportionality and sound administration.

Infringements fall into three main categories:

- Action or omission directly contrary to a specific provision of the legislation.
- Action or omission in breach of these general principles, in particular the principles of transparency and equal treatment and much of the difficulty experienced by authorities in implementing this regime is encountered in dealing with those two principles as worked out in the case law.
- While purchasing authorities have a wide margin of discretion in the decision-making process which they undertake in the context of this regime, that discretion is not unlimited and decisions that plainly involve a manifest error may not be upheld, or at least are susceptible to being set aside upon challenge.

Part II

Procedures

The public procurement legislation provides for an array of procedures: open, restricted, negotiated (with or without competition), competitive dialogue and other newer more complex procedures. In fact, much of the complexity can be stripped out in most procurements by realising that there are essentially three core types of procurement. First, there are procurements conducted in which there is no public tendering whatsoever. These are permitted only in rare cases. Second, there are procurements in which any entity that expresses interest in the process is permitted to bid. Third, there are tenders in which there are two separate stages – first the purchasing authority invites expressions of interest and then, on the basis of those expressions of interest, decides which entity will be invited to bid, fixes those bids, evaluates those bids and awards the contract on the basis of the procedure that ensues.

The essence of all of these procedures is publicity and the opening up of competition. The starting point of any regulatory procurement is the publication in the *Official Journal of the European Union* (OJEU) of a notice – often called the OJEU notice. In each of these procedures, there are two distinct stages. At the first stage, tenderers are evaluated and a decision is taken as to who is to be permitted or to be invited to tender. There is then the second stage at which tenders are evaluated. Case law requires that these two processes be kept separate and that matters such as the experience of the tenderer can be looked at only in the first phase considering the nature of the tenderer, rather than in evaluating the tender.

Key principles

At a simple level, transparency requires two things. First, the existence of the procurement and its outcome must be publicised. Second, the purchasing authority must say what it intends to do and then do what it said it would do. Transparency does not necessarily involve the imposition upon purchasing authorities of an obligation to comply with each and every line of every invitation to tender that is set out by them. It does not create a contractual obligation or an obligation in any other legal basis where there was not otherwise such legal obligation. However, no material change to the procedure for selecting bidders or awarding contracts may be made once the tender documents are published, unless everyone is notified of the change and has a fair chance of dealing with it in their bid.

Equal treatment requires that comparable situations must not be treated differently and that different situations must not be treated in the same way unless such treatment is objectively justified. This principle is the basis of much procurement litigation as it provides a basis for arguments that different or unfair treatment of a bid is unlawful.

Evaluating tenderers

The evaluation of tenderers itself falls into two distinct phases. First, inappropriate tenderers are excluded and then the decision is made as to which tenderers are to be invited to submit a tender in those procedures in which the purchasing authority does not contemplate that all would-be bidders will in fact be invited to bid.

Under the relevant legislation some bidders must be excluded from bidding. This includes, for example, where the would-be tenderer has been convicted of an offence under section 1 or 6 of the Bribery Act 2010, or a range of other specified offences. In addition, a contracting authority may exclude an entity that has committed various other offences or is in default in various other ways, such as has committed an act of grave misconduct in the course of his business or profession, has shown deficient performance in the past or can be shown to have been involved in anti-competitive conduct. In any procedure, the authority may also impose minimum requirements as to economic and financial standing or technical or professional ability and exclude those that do not meet that requirement. At the second stage of this tenderer evaluation, purchasing authorities may operate their own procedures for deciding which bidder will in fact be invited to bid. Transparent criteria must be set out for the operation of such a procedure.

This process of tenderer evaluation is becoming increasingly contentious, especially since the government has sought to exclude tenderers on the basis of past poor performance or default in tax compliance. However, the determination as to whether a bidder is to be regarded as in default of their obligations or to have somehow fallen short of the requisite standard to be treated as an appropriate tenderer is highly controversial and itself open to some political interference.

Evaluating tenders

Once the tenderers have been identified and have been invited to bid, they are provided with the requisite information against which to provide that bid. The title, nature and number of these documents will depend upon the nature and detail of the procedure. In some procedures, there will be a single document requesting the information, in others such as competitive dialogue, there may be a series of tender documents setting out ever more involved or refined ideas as to the requisite requirements of the authority. The legislation requires that the contract must be awarded on the basis of the offer which either: (i) is the most economically advantageous from the point of the contracting authority or (ii) offers the lowest price. In reality, the lowest price criteria is hardly ever used in the construction sector, despite what is sometimes suggested by commentators. Much confusion, though, is caused by the use of nonfinancial criteria which are often ill suited to clear objective analysis.

The contracting authority is required to identify in advance the criteria against which it will evaluate the relative economic advantage of one tender against another, and some of these criteria are identified in a non-exhaustive list in the Regulations. The Regulations state that where a contracting authority intends to award a public contract on the basis of the offer that is the most economically advantageous, it shall state the weighting that it gives to each of the criteria chosen or, in the case of a competitive dialogue, in the descriptive document. Alternatively, it may give the weightings a range and specify a minimum and maximum weighting where it is considered appropriate in view of the subject matter of the contract. Where it is not possible to provide weightings for the criteria on objective grounds, the contracting authority shall indicate the criteria in descending order of importance in the contract notice, contract documents or, in the case of a competitive dialogue, in the descriptive document. Contracting authorities only very rarely take advantage of the possible flexibility in stating weighting and in reality the almost universal practice in the United Kingdom is to set up complicated quantitative weighting systems in which each bid is evaluated by reference to a series of criteria or sub-criteria by a group of evaluators who provide a quantitative output. That quantitative output is then accumulated with other assessments and put into an algorithm that produces a numerical answer determining the most economically advantageous tender. This has perhaps the rather surprising result that the decision as to which is the most economically advantageous tender is very rarely taken by those whose decision it is to take. Rather, their hands are usually tied by the output of groups of evaluators who may not have had any overall view of the bids but have simply looked at the bid by reference to one or more of the separate criteria. The overall decision makers very rarely are in a position to make any choice whatsoever.

Much therefore turns on the selection of the criteria and the weighting applied to them. Not surprisingly, government and other purchasers often see considerable possibilities in using their power to select or craft those criteria in influencing the outputs of the procurement. These criteria must be connected to the subject matter of the contract. This constraint is important in limiting the ability of purchasers, for example, to apply environmental or social criteria with broader impact beyond the instant contract. Thus, it is probably lawful to apply environmental criteria when relating to the quality of the products that are being purchased. It is probably not lawful to apply criteria that seek to impose obligations upon a bidder regarding the environmental performance of all its products. Social criteria are particularly problematic as they have often been seen to be a potential basis for national discrimination and as already explained, that is, the cardinal sin of EU public procurement law. Plainly a criterion that gives preference to an entity because of its past or existing employment of local staff may be discriminatory, but arguably criteria that seek only to give preference to those who undertake to develop local employment once the contract is awarded may not be discriminatory as long as the criteria is applied in a way that any bidder from any EU State could meet the requirement.

Framework agreements

A recurring issue in the construction industry is the tendency of purchasers to use the procedure provided for in the legislation to rely upon framework agreements. These are agreements of different types, which share the basic concept that an entity or entities are selected to be available to enter into contracts with one or more public authorities over succeeding months or years on terms fixed within the tender. There has been considerable litigation concerning these framework agreements, not least because the effect of these framework agreements can be to aggregate demand over a considerable part of the economy or over a considerable geographical area with the effect that using the framework agreement can exclude bidders, often SMEs, from a substantial part of the available business in its sector.

Contract change

Where substantial modification is made to a contract or concluded contract, the purchasing body may be required to terminate the contract and commence a procurement for a fresh contract. Change which is made pursuant to an agreed variation clause so that change is priced ought not to have such consequence but many other changes do. The authorities are rather unsatisfactory in the lack of detail they currently provide as to what constitutes substantial modification for this purpose.

Cancellation of the process

Purchasing authorities have a broad discretion to cancel procurements, but this is not unlimited. A procurement cannot, for instance, be cancelled simply to avoid contracting with winning bidders. There may be other circumstance in which such claims succeed where the decision to cancel can be shown to be in some sense improper or unfair. The law is at an early stage of development on this topic.

Information obligations debrief and disclosure

Once the decision has been taken to award the contract, the Regulations set out a detailed regime under which all those who submitted bids are informed of the outcome and provided information regarding the outcome. That information is expressly intended by the legislator as explained in the case law to be the information that enables dissatisfied parties to bring challenges if appropriate. The theory is that the challenge or threat of challenge will keep the purchasing authority honest. The Regulations provide that information should be given to the loser bidder immediately and that the contract may not be entered into until the expiry of ten days from the provision of the requisite information. That information should include not only any scores awarded to the winner, but also the reasons for the award decision, including the characteristics and relative advantages of the successful tender. Conclusion of the contract before the expiry of the standstill period will likely lead to the contract itself being declared ineffective, as to which further below.

In fact authorities are likely to be subject to more extensive obligations even than those provided for under the regulations. Many public purchasers will of course be subject to freedom of information obligations. These can sometimes be a useful means by which challengers can obtain information, albeit that the timetable for providing information under that legislation is longer than is generally regarded as useful within the context of procurement challenges. Where a procurement challenge is threatened in court, the availability of court procedures for provision of either disclosure before the commencement of litigation or at an early stage of litigation also creates the incentive for authorities often to pre-empt any application by providing that level of disclosure that would otherwise be provided or ordered pursuant to the court order.

The obligation of transparency in procurement, the general government policy towards transparency and, the need for disclosure in litigation are all hard to reconcile with the fact that much information provided by bidders will be commercially confidential. Furthermore if in due course it is decided to seek a re-bid of some or all of the procurement in order to address some issue raised in the challenge, it may be very difficult to do this if key parts of one or other bids have already been disclosed in such a way that either competitors or others have in fact already obtained the requisite information. In litigation, this is often addressed by establishing confidentiality rings in which a large part of the disclosure made in the course of the challenge is made only to lawyers and experts and kept away from those within the client organisation who may subsequently find knowledge of competitors' intentions to be useful. This is not however a straightforward process and creates a particular burden for lawyers and experts involved.

Commencing proceedings

The most usual means of challenging a decision in public procurement is by means of High Court action commenced, similar to any other private claim, by way of claim form. The claim form can be issued in any division of the High Court but cannot be issued in a county court. Proceedings are not normally issued in the Administrative Court. If a claim can be made by way of private law, the Administrative Court ought not to admit a claim as a judicial review claim, albeit that there are some recent instances where that position has been somewhat softened. The limitation period for bringing these proceedings is extremely short. The claim must be brought within 30 days of the date when the economic operator first knew or ought to have known that grounds for starting the proceedings had arisen. The standard of knowledge required to start time running has been decided by the Court of Appeal to be knowledge of the facts that apparently clearly indicate, though they need not absolutely prove an infringement. This is, therefore, a very short limitation period, indeed shorter than almost any other limitation period in the private law arena. The limitation period is subject to certain further qualifications, which may in some circumstances extend the time for bringing proceedings to ten days from the date upon which the relevant decision is taken. Any proceedings started must proceed with a degree of urgency and the claim form must be served within seven days of the date of issue. Once a claim form has been issued with respect to a decision to award the contract and the contracting authority has been told of that issue, the contracting authority is required to refrain from entering into that contract.

Remedies

The first and often most important remedy now lies immediately within the hands of the challenger given that immediately upon issuing the claim form and informing the authority of that claim, the effect of an interim injunction preventing conclusion of the contract is achieved. It is then for the authority to apply to court to have that suspensive effect lifted by order of the court. Authorities have generally been successful in having that suspended effect lifted not least because the final remedies available to the authority include not just injunctions and similar orders, but also the award of damages. That award may itself include an award of damages for the loss of profit or loss of chance of earning a profit providing a contract. Even though the standstill may be lifted, the fact that the automatic standstill is in place often leads to a reconsideration of the situation by the purchasing authority. Challenging an award decision and maintaining a suspensive remains a risky course for the challenger given the costs involved in sustaining High Court litigation. It is likely that in order to maintain a suspension in place, the challenger will have to give an undertaking to the court to meet any losses suffered by the purchasing authority or other affected third parties in the event that it turns out that the claim was misplaced.

The draconian remedy of a declaration of ineffectiveness is also provided for by the Regulations. Where the contracting authority has entered into a contract without any public advertisement and without appropriate legal justification, or where it has entered into a contract without allowing the statutory standstill period to expire or it has entered into a contract despite the fact that it was subject to the automatic suspensive effective referred to above, the challenger may make an application for the contract to be declared ineffective. If the grounds for ineffectiveness are met, the court has no discretion but to make that order unless there are overriding reasons relating to the general interest that require that the effects of the contract should be maintained. An authority, which is the subject of an order of ineffectiveness, must also be the subject of an appropriate dissuasive penalty payment. Where for exceptional reasons identified above, the authority avoids ineffectiveness, the court must consider at least shortening the term of the contract. A contract declared to be ineffective is prospectively but not retrospectively void and it is for the court to work through the consequences of that prospective ineffectiveness. The court is required to have regard to the terms of any agreement made by the parties to the contract before the declaration of ineffectiveness, unless the provisions of such an agreement are incompatible with the overall requirements of the declaration of ineffectiveness and the underlying legal requirements.

Complaints to the EU commission and other challenge procedures

There are a range of other means of challenging procurement procedures. Challenges can be made to the EU commission which can then decide to bring proceedings against the UK Government under Article 258 of the Treaty for European Union. This has the advantage for the challenger of being relatively cost free but it is an unpredictable route of challenge and in reality has only rarely been of much value to the actual challenger because the Commission tends to pursue the case with an eye to its policy goals rather than in the interests of the challenger.

The UK Government has also established a mystery shopper scheme, which enables dissatisfied tenderers to raise their complaints with the Government Procurement Service, which takes the matters up with the purchasing authority. The intention of this procedure is only to improve future conduct and is not intended to affect the outcome of the tender.

Tendering contracts

On the basis of the decision of the Court of Appeal in *Blackpool Aero*, a purchasing authority that advertises for bids is bound by an imputed tendering contact to open and consider bids submitted in accordance with the specified terms. However, it remains unclear when or whether there exists broader contractual obligations governing tenders in the private or public sector.

The JCT has produced a useful Practice Note, Tendering 2012 covering many of the issues dealt with in this chapter.

Notes

1. Arrowsmith, "The purpose of the EU Procurement Directives and: Ends, Means and the Implications for National Regulatory Space for Commercial and Horizontal procurement policies" Cambridge Year Book of European Legal Studies Volume 14 2011–2012.
2. There are a number of value thresholds for service contracts ranging between £100,000 and £175,000. England and Wales share the implementing regulations with Northern Ireland. Scotland has implemented the directive in separate Regulations amended most recently in 2012.

Part III

Preparing for and Inviting Tenders

Chapter 17
Procedure from Brief to Tender

Initial brief

The initial brief from the employer may be little more than a statement of intent, as mentioned in Chapter 2. However, for the project to be successful, the employer's brief must be developed in detail so that due consideration can be given to each of its aspects. The design brief and the design process develop in an iterative manner, with progress on one aspect creating a need for further thought on and consideration of other matters. The more complicated the building, the more important it is that the brief is developed in a systematic way and as early as possible. Full development, certainly on major schemes, is a team affair requiring the preparation of feasibility studies, cost reports and consultations with planning and other relevant authorities.

Developing the brief

Numerous factors need to be taken into account when developing the brief. Some of the main ones are:

- *Employer attitude* – Establish whether the employer requires adherence to strict corporate standards.
- *Environment* – Determine the type and quality of the internal and external environment desired.
- *Operational factors* – Determine the activities the building is to accommodate and the other practical factors that will govern its layout and the relationship of the various elements within it. For example, assess the need to isolate certain items of plant and machinery in order to create acceptable noise and other

The Aqua Group Guide to Procurement, Tendering and Contract Administration, Second Edition.
Edited by Mark Hackett and Gary Statham.
© 2016 John Wiley & Sons, Ltd. Published 2016 by John Wiley & Sons, Ltd.

environmental conditions elsewhere in the building. Also determine the extent of flexibility required in the designed space to provide for future adaptability.

■ *Site* – Identify ownership and any operational hazards. Evaluate access limitations, site topography and soil suitability (using soil tests, trial and bore holes). Survey the site and/or buildings to provide critical information such as access, boundaries, watercourses, trees, existing services, dimensions, levels and condition.

■ *Timescale* – Assess the appropriateness of the desired commencement and completion dates in the context of the practicalities of the situation and other projects that the employer may be planning and determine a programme.

■ *Finance* – Establish the employer's budget and funding arrangements and whether or not the envisaged scheme is likely to be varied if additional funds become available. Determine the relationship between capital and revenue funding and any cash flow constraints.

■ *Market* – Examine the market potential, timing and selling strategies for speculative buildings.

■ *Costs* – Assess overall projected capital and running costs.

■ *Development control* – Determine whether any particular development control considerations affect the site. Liaise as necessary with the development control authority.

Feasibility stage

Developing the design and preliminary cost estimate will lead to a feasibility report, advising the employer whether the project is feasible functionally, technically and financially.

Sketch scheme

The design process usually commences with the sketch scheme. Initially this involves the designer putting on paper preliminary responses to the brief. The sketch scheme will demonstrate the extent of accommodation that can be achieved: for example, the number of new houses on a particular site or the number of shopping units with their associated parking requirements. When the emphasis is on design quality as it would be in a conservation area or in the case of extensions to sensitive buildings, the sketch scheme will need also to address the visual form, style, proportion and materials. Whatever the nature of the drawings or information produced at this stage, the presentation should be capable of being readily understood by those reviewing the proposals.

Unless the employer has dealt with the issue, members of the design team should now advise on development control matters. They must determine whether the proposed development will require development control approval. If not, it is wise to obtain written confirmation of this in case of later dispute. Assuming approval is required, the need to obtain an outline development control approval prior to detail approval should be discussed. This usually becomes relevant when a principle needs to be established, for example a major change of use, the development of a site not designated under a local authority plan or the development of a site over and above local density levels. Unless in a sensitive area, only basic plans or Ordnance Survey map extracts are required, thereby limiting the employer's financial commitment.

Costs

At the start of the sketch scheme the quantity surveyor will probably be required to prepare an initial cost estimate; inasmuch as the sketch scheme will be short on detail yet the employer will expect a reasonably reliable estimate of cost, it will take an imaginative quantity surveyor to address the full cost consequences of outline design information. This will invariably be based on a simple cost per square metre calculation and/or a comparison of outline unit costs. As the scheme develops it will be possible for the initial cost estimate to be refined on an element-by-element basis (see Chapter 17) such that the predicted cost may be amended or confirmed. This will establish whether the scheme is likely to be within the employer's budget. If it is not, then appropriate adjustments can be made – these can be done more readily at this stage than later in the design process. Once the scheme is within budget, the refined elemental estimate provides a cost control document. The cost of each element can then be monitored as the design develops and thereby prevented from unintentionally becoming unduly expensive or over-designed in comparison to other elements.

The employer is now in possession of an outline scheme which meets the basic parameters of the brief, together with a cost plan based on known levels of quality. A decision as to whether to proceed can now be made. Beyond this point a commitment to progress the scheme will result in significant financial outlay.

Procurement

The method of choosing the contractor to construct the building must now be determined as this may dictate the manner in which the detailed design is progressed. For example, if a traditional procurement route is being followed, the design team will develop the design, whereas with design and build procurement the design team may only prepare a design brief, the design itself being completed by the contractor. Each procurement option will have a different time, cost and quality scenario (the procurement triangle – see Chapter 4). These three elements may be held in a particular balance as the circumstances dictate. However, one element must always give way to the others and altering any one element will have an effect on the others – see Figure 4.1.

One interpretation of the time, cost and quality priorities which may be derived from three different procurement routes is given in Example 4.1. It must be stressed that the particular circumstances of each project may result in different conclusions as to the procurement route that best satisfies the priorities of time, cost or quality. For example, a high level of provisional sums within a contract could undermine the cost certainty otherwise afforded by a particular procurement route but would enable work to be started sooner if time is a higher priority. Additionally, it should be noted that for ease of presentation the time allocation for building control approval has been omitted from Example 4.1. This element in itself can influence the choice of the procurement route.

It is advisable to apprise the employer of the procurement options, recommending the most appropriate route, detailing how the priorities are protected and outlining all implications. Only when the method of procurement is determined can the pre-contract programme be finalised. This also represents the point in time when the specific services required of each member of the design team can be defined, fee proposals confirmed and agreements finalised.

Detailed design

The development of the sketch scheme into a workable solution will produce the detailed design to the employer's brief and allow it to be submitted for detailed development control approval. The onus is on the design team to maintain the priorities established within the employer's brief and to extract and agree on any further information required from the employer as the scheme progresses. Whatever the scale of the project, members of the design team have a responsibility to use their skills to:

- produce a good solution to meet their employer's brief;
- consider end and/or future users if different from the employer;
- design to meet the anticipated lifespan of the building to avoid early deterioration or costly maintenance;
- include, as far as is reasonably practicable, health and safety considerations in the design; and
- co-ordinate structure, finishes and services into one complete design.

During this development stage it is wise for the designer to hold preliminary meetings with the local authority to assess the likelihood of development control approval, to define any agreements the employer may have to enter into with the authority and to determine the timescale involved. If reaction is unfavourable or objections from others seem likely, the designer owes a duty to the employer to advise accordingly, indicating the possible extent of negotiation that may be required to achieve approval or the implications of appealing with regard to time, cost and risk of refusal.

In submitting a detailed development control application the design team is likely to have to produce the following information:

- floor layout plans;
- typical sections showing proposed heights;
- all elevations;
- elevations or pictorial views, where relevant, showing the proposals in context with adjoining buildings;
- site plans showing relationship of the proposal to other buildings, orientation, access, parking standards and the like; and
- details of materials and colours.

This information should be presented legibly and accurately in view of the different groups, committees, departments and members of the public that may have the right or duty to give their opinion.

If the application is refused or the application is not determined within an agreed timescale or the employer objects to certain conditions imposed in an approval, the applicant has the right to appeal within a time limit. Great care should be taken in advising an employer on these matters, especially in view of the penalties imposed for unreasonable or unsuccessful appeals. The employer may be well advised to seek expert legal advice prior to pursuing an appeal, particularly if the proposal is complex or risky.

Once a detailed approval is granted, an employer can instruct the design team to proceed to the production stage; if such instruction is given before the approval is granted the employer must accept the risk of abortive work.

Programming

When all submissions for statutory and other approvals have been satisfactorily completed and when the final design proposals have been given, the design team can re-examine the outline programme made at the start of the project and amend it in the light of the stage reached, the procurement option chosen and any financing conditions imposed.

The purpose of the detailed pre-contract element of the programme is to set out a sensible and logical sequence of the various pre-contract operations appropriate to the members of the design team and the required level of their input, together with external factors peculiar to the project. The initial bar chart or network can be expanded to show a programme to include:

- design team progress meetings;
- preparing co-ordinated production information;
- obtaining statutory approvals;
- finalising legal agreements for access arrangements, party wall awards and the like;
- negotiating with the service providers;
- receiving and integrating specialist design elements;
- receiving quotations from sub-contractors and suppliers;
- finalising information for the pre-construction information plan;
- preparing contract conditions and preliminaries, including sequencing and contingencies;
- preparing tender pricing documents;
- preparing pre-tender estimate and updated cash flow prediction;
- completion of pre-tender enquiries;
- provision of tender documents;
- receiving, appraising and reporting on tenders;
- assembling contract documents;
- briefing site inspectorate; and
- naming sub-contractors and suppliers.

Part III

In conjunction with these key points, the team should build in a contingency to cover such activities as co-ordination within the team, the inevitable drawing changes during the process and checking completed documentation. For the programme to run smoothly, all team members should be fully aware of the programme in respect of their own role, the interdependence of members of the design team and the effect that delay by one member will have on the others. It is the duty of the project team to keep the employer informed of progress against the programme and to give advice on any factors which may alter the programme and their effect in cost and/or time.

Design team meetings

Chapter 2 refers to the traditional role of the architect as project leader. However, increasingly with larger projects, employers appoint a project manager to co-ordinate the pre-contract and site works on their behalf. A project manager should seek to manage the effective production of all the required information and ensure that the design team is working cohesively towards the common goal. The project manager should be aware, therefore, of the terms of appointment and limits of responsibility of each team member and ensure that they work together to produce an efficient and coherent design.

It is easy for time to be wasted if the members of the design team are not in regular contact. The project manager should hold regular design team meetings to review:

- new and revised information;
- the detailed design;
- the integration of the structural and services elements;
- specialist design input;
- health and safety principles; and
- progress against the programme.

Clear decisions must be made at these meetings. Time spent constantly changing drawings is unproductive and likely to lead to abortive costs. A record of decisions taken and any action required by a member of the design team should be made so that such actions can be checked off at subsequent meetings.

It is important that members of the design team should not lose sight of the fact that their role is to provide a service to the employer. Many professional practices operate strict quality assurance procedures to ensure the maintenance of good practice within their organisation and the provision of quality service to any employer.

Drawings

The most traditional and widespread method of conveying fundamental information for building is the drawing. It is common for the architect to use computer and overlay draughting techniques and to commence by producing basic plans and sections (general arrangements) on which can be superimposed:

- setting-out dimensions;
- structural engineer's designs;
- services engineers' designs;
- specialists' designs; and
- information for development control submissions.

Once setting-out, structure and services requirements have been determined, the design team may then proceed to the production of large-scale details.

The clarity of drawings and other documents is fundamental to the smooth running of any project. Hence the goal of the design team must be to make the project information as simple and as easy to understand as possible. It is suggested that the most effective way of achieving this is to adopt the conventions of Co-ordinated Project Information (CPI) as set out in a guide first published by the Co-ordinating Committee for Project Information in 1987. More recent standards are set out in the code of practice BS 1192:2007 'Collaborative production of architectural, engineering and construction information'. This code of practice establishes a methodology for managing the production, distribution and quality of construction information using a disciplined process for collaboration and a specified naming policy.

Specifications

At the beginning of the pre-contract programme the design team, in conjunction with the employer, will need to make an early selection of the options for the appropriate performance requirements and/or the type and quality of all materials, goods and workmanship necessary to complete the project. When selecting specific materials and goods the design team will have to ensure that they:

- are appropriate to function, exposure and predicted use;
- are available without undue difficulty;
- provide appropriate levels of thermal resistance, sound attenuation and fire resistance;
- comply with relevant health and safety conventions;
- are environmentally friendly and energy efficient in manufacture and use;
- create the required ambience in terms of space, colour and texture;
- are consistent with the employer's policy on sustainability;
- are easy to clean and maintain; and
- are appropriate to the employer's budget.

When translated into written form, the performance requirements and/or the qualitative details of the selected materials, goods and workmanship comprise the specification. Chapter 20 explains in more detail the different types of specification and their function.

Part III

Bills of quantities

The decisions on specification made by the design team and the employer form the basis of the written contract documentation. The method for presenting this information will depend upon the procurement route adopted. One of the more traditional procurement routes will require the production of bills of quantities. This process is one by which the quantity surveyor analyses the drawings and specification and, by following a standard set of rules, translates them into a schedule of quantified, descriptive items building up the constituent parts of the proposed project.

The primary purpose of the bills of quantities is to provide a uniform basis for competitive lump sum tenders and a schedule of rates for pricing variations. Chapter 22 explains in more detail the different types of bills of quantities and their functions.

Specialist sub-contractors and suppliers

The use of certain materials, goods or installations requires specialist knowledge and skills that are likely to be beyond those possessed by the main contractor. When this occurs, the design team will normally approach specialist suppliers or sub-contractors for the provision of design and quotations for the work.

Quality assurance

Throughout all procedures, the quality of the product or service should be at an acceptable level. Quality assurance is a management process designed to provide a high probability that the defined objective of the product or service will be achieved. To achieve quality consistently, an organisation needs to have a suitable management system in place. That system should be capable of applying appropriate management checks throughout all work stages, be it a design service, a product manufacturing process or a building erection process, and of correcting deficiencies if they occur.

ISO 9000 sets out a series of standards for quality management systems overseeing the production of a product or service. ISO 9000 was originally published by the British Standards Institution under the guise of BS 5750, but is now maintained by the International Organisation for Standardisation (ISO).

ISO 9000 is a generic standard that can be applied to any company wishing to create a quality management system, whether it is large or small, for-profit or governmental, whatever the product or service. ISO 9000 standards focus on a number of principles (defined in ISO 9000:2005 and ISO 9004:2009) of a customer focus, sound leadership practices, involvement of people at all levels, a process approach, a systematic approach to management, continual improvement, a factual approach to decision making and mutually beneficial supplier relationships.

A recognised way for a firm to set up its own quality assurance system is to prepare a quality manual covering:

- the overall policy of the firm as regards quality of service;
- policies on such matters as information services, staff training, resource control and documentation; and
- a procedures manual for the preferred methods of running projects.

The principals of the firm should formulate, endorse and evaluate the manual and communicate its contents to all personnel, ensuring full understanding. A firm may then decide whether or not it wishes to seek third-party assessment by a certification body leading to registration under ISO 9001. Such a body will first examine whether the manual prepared by a firm meets all the requirements of the International Standard and then whether the firm is operating in accordance with the quality manual. After registration, further third-party assessments are carried out at regular intervals to ensure that the firm continues to operate in accordance with its quality manual.

Quality assurance on its own will not improve a firm's professional standards: it only provides an auditable record of performance against the firm's stated objectives, which may or may not embody high professional standards. The adoption of quality assurance will promote consistency in performance only at the level set down within the quality manual.

Obtaining tenders

This chapter has discussed the various elements involved in the pre-contract programme and factors important to the smooth running of that programme. It has also examined the different types of information that are used to convey to the contractor the work to be done and to obtain tender prices. In Chapters 18–23 these types of information are discussed in greater detail and Chapter 24 addresses the procedure for obtaining selective tenders.

Part III

Chapter 18
Pre-Contract Cost Control

Simon Rawlinson

Introduction

This chapter focuses on best practice and key processes associated with Pre-Contract Cost Control. Most of the discussion is focused on the new standard for cost planning developed by the RICS, known as the New Rules of Measurement or NRM1 (2nd edition). We review the purpose of pre-contract cost planning and how estimates are relied upon by clients in a range of circumstances. This chapter also includes a discussion on the wider context of pre-contract cost control, and how it aligns with the work of other consultants. The bulk of this chapter focuses on the two main types of pre-contract estimate – the Order of Cost Estimate and Cost Plan – when they are used, how their content develops over time and how they are presented.

The purpose of pre-contract cost control

The overall purpose of pre-contract cost control is to assist the setting of project budgets and to maintain financial control during the development of solutions to meet the client's need. In providing this discipline, effective cost control provides a number of benefits to the client including:

- certainty with respect to the completion of design to the required level of detail required for stage completion;
- certainty around design development and the project team's buy-in to the budget;
- assurance of the best allocation of money across building elements; and
- clear communication of risks and outstanding design issues.

The Aqua Group Guide to Procurement, Tendering and Contract Administration, Second Edition.
Edited by Mark Hackett and Gary Statham.
© 2016 John Wiley & Sons, Ltd. Published 2016 by John Wiley & Sons, Ltd.

Setting the initial budget is critical. It is an industry cliché that the first cost given is the cost that the client will remember. The initial budget also sets the tone for the project, as the most important 'career defining' decisions will have been made using the least amount of information. If the project team are directed to understand that they are working to budget constraints they will act accordingly. However, if the team are allowed to develop a design that cannot be accommodated within the original budget then credibility in the team will be lost and both time and fee will have been wasted.

Estimates and cost plans are typically considered in the context of a construction project – progressively supporting the development of design solutions in response to a brief. Estimates may also be used to support higher risk activities at the early stage of projects, such as business planning, site purchases, regulatory settlements or due diligence in connection with an acquisition. Understanding the purpose of the estimate, the level of accuracy required and the risk items that need to be taken into account are key aspects of successful pre-contract cost control.

Framework for pre-contract estimating

Cost consultants currently entering the building profession are in the fortunate position of having access to a fully documented pre-contract estimating process such as NRM1, which covers order of cost estimating (OCE) and cost planning (CP) for capital building works. Separate volumes cover measurement for Bills of Quantities for capital projects and estimating for maintenance work respectively. Other professional groups including Civil Engineering estimators also have access to measurement rules, such as CESMM4.

NRM1 is a guidance document, produced with the intention that everyone involved in a project – client, project manager, designers and so on – will understand the estimate and the associated information requirement. As a guidance document, it has the status of best practice. RICS members are not required to follow the guidance, but in the event of a dispute over the level of service provided, adherence with the guidance is described by the RICS as an effective means of demonstrating the application of appropriate professional skill and care.

NRM1 reasserts position of the OCE and CP as management tools, driving delivery of a project to meet its brief within the defined cost limit. In doing so, the discipline of meeting the top down cost limit target is fundamental to pre-contract cost planning. Unfortunately, there are many examples of projects where cost management disciplines have not been applied effectively and where the cost plan 'kept the score' – providing a bottom up assessment of how much a project might cost without the surveyor exerting any influence on the outcome. Cost planning under NRM1 is not a passive process and relies on effective team working to enable design solutions to be developed within cost limits.

- 'Top down' estimating describes the process where the design of a building is directed to meet a clearly defined cost target. Good examples include private-sector housing and under the UK Government's 2011 Construction Strategy, projects delivered by contractor-led consortia.

Part III

- 'Bottom up' estimating occurs where there is no alignment between the brief, budget and design concept – where the cost consultant can exert limited influence on design development. Perhaps the best example of this is the Scottish Parliament project, which was subject to major cost overruns. In the subsequent enquiry, it was shown that the cost planning process had been meticulous, but had very little influence over project outcomes.

The estimating process set out in the NRM1 is fully aligned with the wider design process. NRM1 defines the information required to produce an order of cost estimate or cost plan. This information is produced as a constituent part of a project's design deliverables. The NRM1 references stages in the RIBA Outline Plan of Work 2007[1] and the OGC Gateway reviews (Table 18.1).[2]

Table 18.1　Alignment of estimates and cost plans and project stages.

RIBA Work Stages			RICS Cost Estimating, Elemental Cost Planning and Tender Documentation Preparation Stages	OGC Gateways (Applicable to Projects)
Preparation	A	Appraisal	Order of cost estimates (as required to set authorised budget)	
				1. Business Justification
	B	Design Brief		
				2. Delivery Strategy
Design	C	Concept	Formal cost plan 1	
				3A Design Brief and Concept Approval
	D	Design Development	Formal cost plan 2	
	E	Technical Design	Formal cost plan 3 Pre-tender estimate	
				3B Detailed Design Approval
Pre-construction	F	Production Information		
	G	Tender Documentation	Bills of Quantities (Quantified schedule of works) (Quantified work schedules)	

Source: NRM1. Royal Institution of Chartered Surveyors.

The number of formal cost plans that are to be produced should be set out in appointment documentation.

In addition to providing confirmation of adherence to project cost limits, the production of a cost plan facilitates the following additional benefits:

- alignment of project costs to the specification and quantum of work set out in the brief;
- project team discipline focused on designing to well-defined and agreed cost targets; and
- clear articulation of the cost of risk associated with the project through reference to risk registers rather than blanket contingency allowances.

As design development progresses, the cost plan increasingly functions as a control document, confirming that the intended specification can be constructed within cost plan limits, and recording the transfer of allowances from design contingencies and elements. Parallel to the cost planning process, the cost consultant will work with the rest of the project team to ensure that design solutions can be afforded within cost plan limits. This work might involve option studies and cost checks, but will not involve the preparation of a full formal cost plan.

Order of cost estimate

NRM1 defines an Order of Cost Estimate as 'The determination of possible cost of a building(s) early in the design stage in relation to the employer's fundamental requirements. This takes place prior to preparation of full set of working drawings or bill of quantities and forms the initial build-up to the cost planning process.'

An OCE will be produced during RIBA stages A and B, associated with a feasibility study. Stage A is concerned with the identification of project objectives, needs and constraints – enabling a business case to be prepared and tested. Stage B involves the preparation of an initial statement of needs and constraints for incorporation into a design brief.

Limited design information will be available at this stage and an OCE can be produced by one of three methods:

- Floor areas – typically gross internal floor area.
- Functional units.
- Elements – based on defined rules for measurement of each element.

Details of the unit rates and element unit rates that can be used in the preparation of an OCE are described in NRM1. NRM1 also describes the methods used to extract unit rates from cost analyses.

Costs calculated using these methods comprise the works cost estimate, which should also include separately stated allowances for facilitating works and preliminaries.

OCEs prepared on an elemental basis are sometimes described as cost models, emphasising that they are not based on a specific design, but clearly show how costs will be allocated on a project.

Information used to prepare an order of cost estimate

Information that is to be provided by the client and project team is scheduled within the NRM1. A schedule of the information received from the client and project team is scheduled in Table 18.2 below. The table does not set out the full information requirement which is detailed in NRM1 and is reproduced to provide an insight into the volume of information that an estimator might expect to receive to produce a well-founded OCE. The design development programme should allow time for the OCE to be produced once these deliverables are available for use.

Sources of cost information upon which OCEs can be based include the Building Cost Information Service of the RICS, Commercial Price Books and data sources managed by public bodies such as the Cabinet Office.

OCEs are very important because they provide the basis on which project budgets are set. Accordingly, an OCE should include a generous risk allowance for design development. However, if the OCE does not align with the aspiration of the client and the design team at the outset of a project, there is a risk that a costly redesign might be required at a later stage. Chapter 7, discussing Value and Risk Management, makes the point that the costs of design changes and the willingness and ability to make changes decrease as a project proceeds – even into RIBA Stage C. Accordingly, the accurate assessment of the costs of meeting the brief – even at the earliest stage of a project – will make a great contribution to the direction of the team to meet the client's objectives.

Table 18.2 Information requirement for the production of an order of cost estimate.

Client	Architect	Services Engineer	Structural Engineer
Site location	Design study sketches	Indicative services specification	Advice on ground conditions
Statement of building use	Floor area schedule	Indicative environmental strategy	Indicative structural strategy
Statement of floor area	Minimum storey heights	Availability of utility connections	Initial risk register
Initial project brief	Schedule of accommodation	Initial risk register	
Details of enabling works	Indicative specification		
Indicative programme	Indicative environmental strategy		
Project constraints	Initial risk register		

Treatment of on-costs and other costs in order of cost estimates

NRM1 provides a clearly structured and consistent approach to the calculation and presentation of non-construction costs. The intention of this presentation is to facilitate the clear communication of how costs are distributed across projects and to enable more effective benchmarking.

NRM1 provides rules for the calculation of the following on-costs at the OCE stage:

Main contractor's preliminaries	Main contractor preliminaries only. Specialist contractor preliminaries are included within the measured work. Calculated as a percentage addition on the works cost.
Main contractor's overheads and profit	Main contractor overheads and profit only. Sub-contractor overheads and profit included in the measured work. Calculated as a percentage addition on the works cost.
Project/design team fees	Percentage allowance for professional fees. An allowance for fees for pre-construction services by a contractor could be included.
Other development/project costs	Lump sum allowance for planning fees, decanting costs, and so on.

Table 18.1 shows how on-costs are incorporated into the OCE. The estimate inclusive of preliminaries and overheads and profit is termed the works cost estimate. The addition of project/design fees and other project costs gives the base cost estimate.

Other costs that are included within the OCE are risk allowances and inflation. NRM1 requires a more structured approach to the presentation of both. NRM1 recommends that risk allowances are split into four categories:

Design risk	Construction risk	Employer change risk	Other employer risk
Design development	Site conditions	Allowance for employer-driven changes	Acceleration or postponement
Changes in esti- mating data	Ground conditions		Availability of funding
Third party risk	Existing services		Unconventional procurement strategy
Statutory requirements	Delays by statutory undertakers		
Procurement strategy			

Part III

NRM1 advises that the OCE should include four separately calculated allowances. At an early stage, these allowances will be calculated on a percentage basis. However, the rules anticipate that there will be a reference to a formal risk identification process.

Inclusion of risk allowances with the base costs gives the project cost limit, for example, the costs which the client and project team can control. Inflation costs are included as a final line item in the estimate. Separate allowances are given for the following:

- Tender inflation – percentage allowance calculated from the base date to the date of the return of the tender.
- Construction inflation – percentage allowance calculated from the date of the return of the tender to the mid-point of the construction programme.

NRM1 rules allow that some aspects of inflation, such as related to changes in market conditions or exchange rates, are dealt with as separate risks rather than in the inflation allowance, which should be based on long term trends.

The total cost of the OCE is termed the Cost Limit (including inflation).

Presenting an order of cost estimate

Figure 18.1 sets out the presentation of a typical order of cost estimate. With respect to the communication of the basis of the estimate, the key principles are:

- It should be presented as a high level estimate, for example, on a £/m^2 of gross internal floor area or £/functional unit basis.
- It should be presented as an outturn cost estimate – inclusive of risk and inflation allowances.
- It should include a full schedule of exclusions.

The NRM1 rules detail the typical contents of an OCE report and, in addition to the estimate itself, the report should also contain the following:

- Project description.
- Cost statement.
- Project information.
- Area schedule.
- Statement of assumptions.
- Base date.
- Inclusions and exclusions.

The build-up to each constituent element is calculated in accordance with NRM1 rules. The inclusion of the detail of the assessment within the OCE report is at the discretion of the quantity surveyor.

Constituent
Facilitating works estimate
Building works estimate
Main contractor's preliminaries
Sub total
Main contractor's overheads and profit estimate
Works costs estimate
Project/design team fees estimate (if required)
Sub-total
Other development/project costs estimate (if required)
Base cost estimate
Risk allowances estimate
    (a) Design development risks estimate
    (b) Construction risks estimate
    (c) Employer change risks estimate
    (d) Employer other risks estimate

Cost limit (excluding inflation)
Tender inflation estimate
Cost limit (excluding construction inflation)
Construction inflation estimate
Cost limit (including inflation)
VAT assessment

Figure 18.1   Constituents of an order of cost estimate.

Cost plans

Cost plans are produced on projects at key points during design development in RIBA Stages C, D and E. Cost plans develop progressively during the RIBA Stages, and the idea of using a consistent framework of elements and sub-elements is to provide a common frame of reference – tracking change, budget transfers and the mitigation of risk. Design development that occurs during the RIBA stages is as follows:

- Stage C – Concept design, including outline proposals.
- Stage D – Development of the concept design, completion of the brief, detailed planning submission.
- Stage E – Preparation of the technical design in accordance with the brief to enable coordination of elements and components for statutory compliance and safe construction.

The output of a cost plan will be used to confirm that a project is meeting its business case and is offering value for money; it will support the development of practical and balanced design solutions which optimise the allocation

Part III

Table 18.3   Information requirement for the production of cost plan 1.

Client	Architect	Services Engineer	Structural Engineer
Confirmation of authorised budget	Concept design drawings and models	Concept design drawings and schedules	Survey reports including contamination and geotechnical
Confirmation of the brief	Area schedule	Outline specification	Environmental risk assessment
Confirmation of programme including critical events	Outline specification	Design strategy statements, e.g. environmental and sustainability and so on	Concept design drawings
Project specific requirements including phasing	Room data sheets Schedules of fittings, furniture and equipment	Specialist reports	Formation and excavation levels
Authority to proceed to next stage	Design strategy statements, e.g. fire, acoustics, security and so on	Utility connection reinforcement requirements	Outline specification
	Risk register	Risk register	Construction methodology
			Risk register

of expenditure, helping to ensure that cost limits are adhered to. The cost plan also helps to establish the relationship between needs and wants articulated in the design and cost and time consequences, captured in the cost plan and programme. Detailed information requirements are set out for each Cost Plan. An abridged summary of the information requirement for Cost Plan 1 produced at RIBA Stage C is set out in Table 18.3.

Depending upon the client's procurement strategy, a contractual commitment could be made with a contractor at any of the RIBA design stages. Accordingly, the cost plan is a key decision-making document – informing decisions to invest in further design, planning submissions and market engagement.

Cost plans follow an elemental structure defined by NRM1, but their actual content will vary in accordance with the level of detail available. Cost Plan1, produced at RIBA stage C, will use a condensed list of sub-elements and, where design work has not been completed, costs will be calculated on the basis of gross internal floor area. By contrast, at the end of Stage E, NRM1 envisages that Cost Plan 3 will be based on the level of information that will be provided in a full tender. In this instance, the estimate will be based on cost checks carried out on preferred design options as the solution is developed. Although the structure is consistent throughout the process, the level of detail will vary. A key skill of the cost consultant is to provide an appropriate level of analysis and control at each stage.

Cost planning under NRM1 is a progressive process; cost targets are set in Cost Plan 1, representing an efficient allocation of expenditure and an affordable solution to the client's brief. Subsequent issues of the cost plan will be based on better information but should always refer back to the targets, reconciling cost transfers between elements and so on.

The works cost elements are described in a four-level set of tabulated rules. The definition of an element is that it is a major physical part of a building that fulfils a specific function, or functions, irrespective of its design, specification or construction. The classification enables these elements to be broken down into progressively more detail as follows:

Group element	The main headings used to describe the facets of an elemental cost plan.
Element	A major part of a group element. A cost target can be set for each element.
Sub-element	A part of an element. A cost target can be set for sub-elements.
Component	The measured and costed item that forms part of a sub-element or element.

At the component level, the rules define both the measurement unit and measurement rules and also provide details of the labours and other items that are to be included within the scope of the item. Clearly, consistent application of the rules will provide well-founded estimates and will enable consistent comparisons of cost benchmarks.

Figure 18.2 shows how the four-level hierarchy is applied to the Upper Floors Element, including a sample of the measurement rules, demonstrating how NRM1 provides a clear and consistent definition of an item of work dealing with ambiguities in work classification and so on.

Sources of price information for the measured items in Cost Plans include analyses of completed projects, market testing exercises, price books and bottomup estimates. As inflation and construction risk items are costed separately, information derived from projects should be normalised to remove additions for inflation, risk and other on-costs such as design fees.

Treatment of on-costs and other costs in cost plans

NRM1 compliant Cost Plans follow the structure for on-costs and risk established in the OCE, albeit with progressively more detail. NRM1 provides rules for the description and calculation of preliminaries and fees which could be completed with agreed costs from the project as procurement proceeds. For example, a contractor's first stage prices for design fees, preliminaries, overheads and profit could be incorporated into a Cost Plan once agreed.

Where commercial information is being input into the Cost Plan, it is desirable that NRM1compliant data is obtained via the tendering process. For example, the preliminaries section includes 14 cost centres which could be populated

Part III

Group element 2.0: Superstructure
Element 2.2: Upper Floors

Sub-element	Component	Unit	Measurement rules for components	Included	Excluded
2 Balconies Definition: Internal and external balconies which are not an integral part of the upper floor construction.	1 Balconies: details, including floor area (m²) of balcony to be stated.	Nr	C1 Where components are to be enumerated, the number of components is to be stated. C2 Work to existing buildings is to be described and identified separately. C3 Contractor designed work is to be described and identified separately. Note: Where the contractor is only responsible for designing specific elements and or components of the building project (i.e. not the entire building project).	1 Purpose-made balconies, which are not an integral part of the upper floor construction. Comprising bolt-on frame, decking, soffit panels, integral drainage/drainage trays and balustrades/handrails. 2 Protective coatings and paint systems. 3 Surface treatments (e.g. surface hardeners and non-slip inserts). 4 Fittings and fixings. 5 Sundry items. 6 Where works are to be carried out by a subcontractor, subcontractor's preliminaries, design fees, risk allowance, overheads and profit. Note: Allowance to be made within unit rate applied to element or component for subcontractor's preliminaries, design fees, risk allowance, overheads and profit.	1 Proprietary bolt-on balconies (e.g. 'Juliet' balconies – included in sub-element 2.5.5: Subsidiary walls, balustrades, handrails, railings and proprietary balconies). 2 Low level and dwarf walls, balustrades, handrails and railings to external walkways which form an integral part of the building envelope and the like (included in sub-element 2.5.5: Subsidiary walls, balustrades, handrails, railings and proprietary balconies). 3 Drainage to balconies which is not an integral part of the balcony unit (included in sub-element 2.2.3: Drainage to balconies).

Figure 18.2 Upper floors element hierarchy. Source: NRM1, Royal Institution of Chartered Surveyors.

from a contractor's submission, but which are less likely to be fully costed by a commercial cost manager. The treatment of on-costs in Cost Plans can be summarised as follows:

Main contractor's Preliminaries	Progressive rules-based assessment of cost significant items of main contractor's preliminaries only, including consideration of the impact of programme on time related items. Sections include management and staff, site establishment, temporary services and so on.
Overheads and profit	Percentage addition for main contractor's overheads and profit only – potentially based on market testing.
Project/design fees	Separate allowances for professional fees, contractor's pre-contract services fees and contractor's own design fees – potentially based on market testing.

Other costs that are included within the Cost Plan include a separate section dealing with other costs. Depending on the project and the client's reporting requirement, these could include land acquisition costs, finance charges, fees (planning, building control, etc.), insurances and so on.

In common with the OCE, there are also standalone sections for risk allowances and inflation. The assessment of other costs will become better defined as the project progresses. With respect to risk, as the design develops and risks are mitigated, transfers will take place from the risk budget into the base cost. It is essential that these transfers are recorded so that it can be demonstrated that contingency expenditure has in fact reduced risk exposure rather than simply being lost to scope creep. Whilst NRM1 strongly recommends the linking of the cost plan to a formal risk management process, including links to risk registers, there is no specific requirement to use quantitative risk assessments.

Inflation costs are included as a final line item in the Cost Plan and are calculated using similar methods to the OCE. These are repeated here for ease of reference. Separate allowances are given for the following:

- Tender inflation – percentage allowance calculated from the base date to the date of the return of the tender.
- Construction inflation – percentage allowance calculated from the date of the return of the tender to the mid-point of the construction programme.

Presenting a cost plan

Figure 18.3 sets out the presentation of the summary page of a typical Cost Plan. With respect to the communication of the basis of the estimate, the key principles are that it should:

- be presented as an analysis of the cost of the current level of design development;

Part III

- reference the cost limits established at the outset of the project;
- be presented as an outturn cost estimate, inclusive of risk and inflation allowances;
- reconcile major changes since the last cost plan/estimate; and
- include a full schedule of exclusions.

The NRM1 rules detail the typical contents of a Cost Plan. In addition to the items described as accompanying an OCE report, the following additional information should be provided:

- Details of the information and specification on which the Cost Plan is based.
- Reconciliation of changes to cost limits, including reasons.
- Value engineering options.
- Cash flow forecast.

The Cost Plan may also include sections dealing with VAT, Capital Allowances plus Grants, as well as Allowances available for Land Remediation. However,

Part III

Constituent
Facilitating works estimate
Building works estimate
Main contractor's preliminaries
Sub total
Main contractor's overheads and profit estimate
Works costs estimate
Project/design team fees estimate
    (a) Consultants' fees
    (b) Main contractor's pre-construction fee estimate
    (c) Main contractor's design fee estimate (if applicable)

Sub-total
Other development/project costs estimate (if required)
Base cost estimate
Risk allowances estimate
    (a) Design development risks estimate
    (b) Construction risks estimate
    (c) Employer change risks estimate
    (d) Employer other risks estimate

Cost limit (excluding inflation)
Tender inflation estimate
Cost limit (excluding construction inflation)
Construction inflation estimate
Cost limit (including inflation)
VAT assessment

Figure 18.3   Constituents of a cost plan.

these are specialist areas and the content for these sections may need to be provided by other consultants.

The typical layout of a Cost Plan summary will follow a similar format to the OCE summary, albeit with additional detail for fees and other development costs. Clearly the details sitting behind the summary will be much more extensive.

Challenges associated with the production of cost plans

Issues that need to be considered in the production of NRM1 compliant cost plans include:

- Management of resources to complete the estimate. NRM1 sets out a very detailed requirement which should be applied proportionately during the design development process. The rules provide plenty of scope for discretion in their application and also encourage the re-use of cost checks and other information where this continues to reflect design development.
- Time to complete the estimate. Given the level of detail required in the tabular rules, a substantial resource will be needed to complete a full cost plan. The overall design programme will need to allow a period of time for cost plan production following the substantive completion of design development. Designers tend to work right up to the end of a stage, so the programming of design deliverables and reuse of cost check outputs will help to ensure that cost plan is produced in a timely manner – and is aligned with the development of the design.
- Availability of information. The schedules of information requirement set out in NRM1 are very comprehensive and it is conceivable that not all of the information will be made available prior to the completion of a cost plan. It is, therefore, good practice to record the information used to prepare an estimate, either to communicate risks associated with 'unknowns' or to capture the impact of late-issued information.
- Effective communication. Cost plans are large documents intended to provide management control. As a result, it can sometimes be difficult to articulate to the client the key issues or decisions that need to be made through a Cost Plan report. Executive Summaries and supplementary reports dealing with specific issues may be necessary for clear communication and timely action.
- Adjusting benchmark costs from projects or market testing for inflation and risk allowances.
- Keeping risk allowances separate. As well as adjusting project data for risk allowances, RICS also recommend that design and tender benchmarks are more appropriate than data sourced from final accounts.

Cash flow

The capital cost of a building project will typically be met by a combination of the client's funds and project finance sourced from financial institutions. Clients are

Part III

keen to minimise their funding costs and do so by agreeing for a drawdown schedule of funding with their lenders. A project cashflow will provide the information
required by the client to agree on the funding schedule with their lender.

The cashflow may also need to reflect other payments incurred during the
project including:

- Site purchase costs.
- Construction costs.
- Costs for advance purchase of materials or pre-construction services.
- Utility connections.
- Planning and building control fees.
- Commuted payments, as part of planning agreements.
- Professional fees.
- Insurances, arrangement fees and other charges.

A cashflow forecast should be prepared to reflect the withholding of retention
funds during the construction contract.

Pre-contract cashflows should be prepared with reference to the planned construction programme, reflecting the critical path of the project and the profile
of spending during different phases. Once a contractor is appointed, either for
pre-contract services or for the main contract, the cashflow should be updated to
reflect the contract sum, dates of possession and completion and the contractor's
programme.

Whole life costs

Buildings are long lived assets, and often have quite high operational costs, related
to heating, ventilation and lighting, repair maintenance and so on. Research undertaken by CeBe (Constructing Excellence in the Built Environment) identified that
the cost of maintenance over a building's lifetime can often equate to at least three
times the original construction cost.

Given the importance of operational efficiency, it is surprising that whole life
costing has not been adopted more widely, but with the sustainability agenda very
high on many clients' agendas, whole life costing commissions are becoming more
common. The issue of sustainability is more broadly addressed in Chapter 33 in
this book.

The whole life cost agenda is potentially a very powerful tool for project teams,
helping to create the case for investment in solutions which promote greater productivity, flexibility, durability or longer operational life. However, as with all
forms of analysis, a whole life cost assessment needs to be prepared in a way which
meets the client's objectives and which provides an accurate representation of
future performance. The challenges affecting the take-up of whole life cost analysis
include difficulties in obtaining unambiguous and corroborated performance and
durability data, together with some confusion surrounding the use of discounted
cashflows.

Without getting into all the details of the production of whole life cost estimates, quantity surveyors should be aware of the following considerations:

The purpose of the assessment – whole life cost studies are typically produced for one of three purposes:

- An estimate of the operational cost of an asset. In this instance, the whole life cost study should cover all potential sources of cost associated with the occupation of a building. These cost centres might include energy, cleaning, insurances, maintenance and so on. Total operating cost assessments such as these are typically used by clients to confirm that they will be able to afford to run and maintain their assets.
- Capital asset replacement. Whole life cost studies which focus solely on modelling the operational life of durable assets such as mechanical systems and roof finishes that require replacement during the life of the building, enabling the client to plan for long-term maintenance obligations, such as the pricing of a sinking fund.
- Option comparisons. Option studies can be prepared to identify preferred options on the basis of long- term performance. The approach could be used to select alternatives that have either different energy use, maintenance or replacement profiles such as window systems, floor finishes, and air conditioning options. Option studies involve the comparison of alternatives which might have different life spans, replacement cycles, income or expenditure profiles. As a result, in many instances, comparison can only be undertaken using discounted cash flow techniques. The use of discounting enables cash flows which occur in different time frames to be totalled and compared on a like for like basis, enabling a best value option to be selected on the basis of net present cost/value.

When reviewing the results of option studies based on life cycle cost methodologies, project teams should ensure that the following aspects of the study have been properly taken into account:

- Discount rate. The selection of the discount rate should take into account the client's requirements. In the case of the public sector, discount rates are published by the Treasury. In the private sector, discount rates generally reflect the client's cost of finance or expectations for rates of return. As the discount rate can have a big impact on the end result of an assessment, its value must always be confirmed by the client.
- Consistent calculation of costs for all options. Where appropriate, both cost and income streams associated with each option should be considered. For example, all cleaning and maintenance costs should be included in a floor finishes assessment, together with some revenues associated with the disposal of high value, long-life assets such as stone finishes.
- Presentation. The presentation of a whole life cost study based on discounted cash flow should make it clear that the reported cost does not reflect what the client or end user will actually pay. Furthermore, if the net present cost differential between two or more options is relatively small, the team should ensure that other criteria, such as the initial capital expenditure, are considered in the selection of the preferred option.

Part III

Summary

Pre-contract cost estimating plays an essential role in helping the project team meet the client's requirements in a cost effective way. In the current financial climate, where public sector clients are facing reduced budgets, and where private sector clients cannot rely on the appreciation of capital values to compensate for cost overrun, cost is really important, and it could be argued that the need for highly skilled cost advice has never been greater. However, to be effective, cost consultants need to work with their project teams to develop well-targeted solutions at an early stage – avoiding any risk of abortive design or delay that might result from an overly expensive design solution. This is not a case of accommodating scope creep without challenge, but in getting the whole team to engage with and own the budget.

Successful pre-contract cost control provides clients with certainty around project outcomes whilst providing design teams with a clear sense of direction. In doing so, communication and teamwork skills aimed at getting the project team to deliver a design within the budget are as much core skills as estimating and benchmarking.

Notes

1. NRM aligns to the RIBA outline Plan of Work 2007 and which has subsequently been superseded by the RIBA Plan of Work 2013. Accordingly, and for the purposes of this Chapter, the RIBA Outline Plan of Work 2007 is referenced in Table 18.1.
2. OGC Gateways form part of a formal project review process developed by the Office of Government Commerce (OGC). OGC Gateways are referenced in NRM1. Information on OGC Gateways can be found on the Cabinet Office website.

Chapter 19
Drawings and Schedules

Peter Ullathorne

The language of drawing

The act of drawing is to use the language of drawings to express thoughts, creative and technical work in a graphical manner and to engage with a wide range of audiences, society and cultures. Drawings help people to see and understand ideas that would be inappropriate, or at any rate ineffective, to explain in words. Drawings transcend the confines of written and spoken languages and become a 'lingua franca' that crosses geo-political boundaries and cultures. The act of drawing develops thinking, invention and invariably leads to action. Drawings used in the worlds of architecture, planning, engineering, construction and facility management have developed in line with advances in these disciplines. Drawings are a form of surrogate reality before reality itself. No longer do we employ draughtsmen to translate sketches into accurate depictions and scalable drawings. No longer do we create perspectives by using vanishing points and colour them using methods developed by Jasper Salwey.[1] We have handed the production of accurate drawings and three-dimensional representations to computer-based technologies using highly complex software. Such software will also convert hard-edged drawings back into sketch form.

The changing role of drawings and documents

Design team members (with their employers) must collaborate effectively to create viable buildings. Successive and influential reports by Sir Michael Latham[2] in 1994 ('Constructing the Team') and by Sir John Egan[3] in 1998 ('Rethinking Construction') identified the shortcomings of then current practice, the need for effective partnering and the use of co-ordinated project information as a

The Aqua Group Guide to Procurement, Tendering and Contract Administration, Second Edition.
Edited by Mark Hackett and Gary Statham.
© 2016 John Wiley & Sons, Ltd. Published 2016 by John Wiley & Sons, Ltd.

contractual requirement. Egan recommended integrated project processes and teams, recognising the potential for IT to play an important part in achieving his and Latham's ambitions. Between then and now, the technique of Building Information Modelling (BIM) has developed to such a degree of competence that, in 2011, the Government's Chief Construction Adviser Paul Morrell called for all Government projects over £5M to be designed using BIM. This important technology is described in Chapter 21.

With BIM we now have a considerably more useful version of the volumetric model used by the master masons of old, using advanced software and hardware. These models are put together by all disciplines, can be tested and are powerful simulacra which are accessible to stakeholders and regulators, as well as design teams and contractors. The creation of such powerful modelling techniques reduces the risks associated with compiling viable briefs and in the ensuing design process which inevitably produces prototypes. Prototype risks are reduced significantly comparing favourably with car and aircraft design techniques.

Project risks for the employer, the architect and the contractor are reduced by the architect and design team to achieve a set level of quality in the design documentation and by all parties using quality management processes. Above all, those who produce project information, when preparing their work, have to think carefully about the purpose of the particular set of drawings or documents and the users of the data. For example, drawings and a description of a project aimed at securing outline planning consent will be only of passing interest to a contractor at the tender stage.

In history, drawings achieved extraordinary importance when they allowed the design function of the master mason to become detached. Drawings became the designer's mute representative on site, allowing the architect to be based remotely and work on a number of projects. The architect's role became defined independently of the craftsman and allowed unrestricted development. No longer was the architect obliged to be the combined designer/craftsman, based for long periods of time on a particular site. However, no longer could the architect provide constant design development, interpretation and experimentation on site on a day-by-day basis as the work progressed. Drawings produced an imperative to have ideas and technical solutions well thought and developed before issue. That imperative continues.

The general principles of architectural drawing have been developed over centuries and today's architects have inherited a long tradition. With developing drawing techniques they have striven always to communicate intent with increasing accuracy and completeness. The triadic system of plans, sections and elevations in use today was developed before the 13th century. Scalar drawing emerged around the 15th century, with constructed perspective drawings coming into use in the 16th century. The 17th century saw the development of projections, stereometry and descriptive geometry, followed in the 18th century by the codification of the triadic system, sciography and rendered surfaces. Axonometrics made an appearance in the 19th century. What of the 20th and 21st centuries? The need for even greater accuracy of project representation, combined with the powerful calculation and representational powers of computing, has seen the strident development of computer aided design (CAD) technologies. The ability to create complex data models of complex building forms, and further to assess their performance in terms of design experience, quality of internal environment, environmental impact and stability under extremes of weather and seismic conditions has transformed design practice. Employers and users can now test potential building designs before committing to project proposals. Statutory authorities can also test projects against their own parameters. The ability to create highly realistic models has significantly reduced the risk of project uncertainty. Performance is predictable. BIM has arrived and, at the time of writing, the Government Construction Strategy (GCS) requires that the Government will require all centrally procured Government projects to be fully collaborative 3D BIM (Level 2 – with all project and asset information, documentation and data being electronic) as a minimum by 2016. Essentially, Level 2 is a managed 3D environment comprising design files for separate disciplines with data. Enterprise resource planning software manages and integrates commercial data with the 3D files. At Level 2, construction sequencing (4D) and cost information (5D) may be added. BIM is covered in detail in Chapter 21. However CAD has brought its own problems. The software used to create designs using CAD/BIM is proprietary, highly complex and impenetrable to all but software experts. The question of an architect's liability for work done using CAD/BIM remains under discussion, although a key consideration is that the architect remains liable for his work, including the accuracy of drawings, schedules and calculations, whether produced by pencil or computer. The need for effective face-to-face communications within teams using BIM was reinforced in an anonymous landmark claim in the USA settled in May 2011. Placing an over-reliance on natural BIM communications proved expensive when the services layout that worked in the model did not work on site. As a result, insurers are now requiring designers to have nothing to do with defining the means, methods, safety and sequencing of works. This conflicts with the requirements of CDM which is a matter that remains to be resolved.

Drawings and documents are not products. One issue is that repeated and uncontrolled use of design documents by an employer may cause the documents to be treated as 'products' generated by the designer and may give rise to product liability risks. There are also issues concerned with the ownership, use and transmission of design data (or 'instruments of professional service') which could include contractors using CAD/BIM data never intended for that purpose, the integration of CAD/BIM data into facilities management programmes or the reuse of electronic project data in projects far beyond the original purpose or well out of time.

Services such as 'delayering' information, using CAD/BIM data as the basis for an 'as built' survey set of documents, or allowing the data to be used as the basis for a facilities management programme or operations, may be regarded as additional services beyond the scope of the standard fee. The attitude currently prevailing is that electronic information is a component of the instruments of service and is only for the employer's benefit on a specific project and for a specific use in relation thereto. Because of the rapidity of development in software and the hardware technologies used, designers need to be indemnified against any future issues arising out of compatibility, usability or readability of design information that is put to uses for which it was not intended.

Quality

Drawings represent the most important means by which designers convey their intention to the rest of the project team and to statutory authorities. As stated in Chapter 17, the clarity and accuracy of drawings is fundamental to the smooth running of any project. This cannot be over-emphasised. Clear, concise, well-planned and co-ordinated drawings not only make the information they contain easy to understand but they also inspire confidence. Conversely, poor drawings do little except reveal the designer's lack of knowledge and inability to conduct affairs in a well-ordered manner.

Standards

The quality of information produced by the architect will either help or hinder the defence in any possible future legal action against him or for which he has to produce information for the defence of others. Recommendations on the preparation of drawings, especially working drawings, are contained in BS1192:2007 'Collaborative Production of Architectural Engineering and Construction Information': and BS8888:2011 for technical product information which is more appropriate for the design of products. BS88541-2:2011 defines a library of objects and 2D symbols to be used in architecture drawings, planning and modelling. BS1192 is relevant to the use of BIM techniques. Also ISO 13567 (1-3) 1998-9 is an international CAD/BIM layer standard applying to technical product documentation, organising and naming layers for CAD/BIM in three parts. It became applicable fully in 1999.

The method of producing drawings will vary from office to office depending upon the size and resources of the firm concerned. Many rely heavily on the use of CAD and related BIM techniques, some others on traditional draughting techniques, but most will use a combination of the two depending on the stage of the project. However, even at the earliest stages of design, software such as 'SketchUp' will augment the processes of creativity and communication. The type of employer, the nature of the project, the degree of involvement with other consultants and their resources and, finally, the programme for the work will all determine how the drawings at each stage are to be dealt with.

But if quality is to be assured, there are simple rules that need to be observed, no matter how the drawings are produced.

Quality manuals

An increasing number of professional firms have written quality manuals used to guide and control the quality and process of preparing documentation and conducting their professional and technical affairs. Quality manuals need to be assessed in action over a period of time by an independent organisation before the latest international standard, BS EN ISO 9001:2008, can be awarded. This accreditation provides employers with a degree of assurance that the firm follows specific procedures in the way that it deals with employer assignments. An increasing number of employers in both the public and private sectors require that firms indicate they are quality assured when they make proposals for work.

Quality manuals help firms to:

- provide an imaginative, competent and consistent professional service to their employers;
- achieve, sustain and improve the quality of design services provided in a way that will consistently meet the employer's requirements;
- provide the documented assurance to employers that the intended quality of service has been and will be achieved; and
- provide an authoritative standard against which actual performance can be assessed in the unfortunate event of any dispute in which the designer is involved.

Quality procedure codes

Quality procedures call for documents to be coded. An example is a four-part code comprising:

- **Document code**   To define the document type. These could be:
  - MA Manual.
  - PR Procedure.
  - FM Form.
  - PN Practice note.
  - RR Register (for system control).
- **Responsibility code**   To define the area of activity within the quality management system and those who are responsible for the writing, approving and distribution of system documents. These could be:
  - AD Administration.
  - DM Design management and project control.
  - SM System management and maintenance.
  - MS Management system information.
- **Document numbering code**   To identify the serial number of the document within its own document type and responsibility code series. These can be simple two-digit numeric references such as 01.
- **Status code**   To identify the revision status of specific documents. It also distinguishes between documents that are still in draft format and those that have been brought into use. Examples could be:
  - OA First draft.
  - OB Second draft et seq.
  - 01 First issue.
  - 02 Second issue et seq.

For example: PR/DM/01/01 would denote the first issue of document 01 related to procedures for design management.

Part III

Quality review

The procedures described below need to be established in order to review, audit, initiate both corrective and preventive action, provide internal feedback and record employer satisfaction (or dissatisfaction):

■ **Management review**   Reviews of the quality system in the firm should be conducted by the quality manager at least twice yearly to evaluate its effectiveness and provide for continual improvement.
■ **Internal audit**   All processes within the firm should be subject to systematic and independent internal audit by suitably trained staff in accordance with audit programmes.
■ **Corrective and preventive action**   The cause of all non-conformity with procedures and standards is identified through:
    – risk assessment;
    – audit;
    – design review and verification;
    – feedback; and
    – post-project review on selected projects.

The results should be analysed and appropriate corrective action implemented to prevent recurrence.

■ **Internal feedback**   A simple mechanism should be established to enable all staff to propose changes and improvements to the quality system based on experience. All information and proposals should be considered and action taken recorded.
■ **Employer satisfaction (or dissatisfaction)**   The level of employer satisfaction should be recorded on both ongoing and completed projects. This is effected through internal audit during the progress of projects and post-project review on completed projects.

Types, sizes and layout of drawings

The types of drawings vary as the project proceeds through its programme, as described later in this chapter. However, for efficient and economic use of time, all drawings should:

■ be on standard-sized sheets laid out in such a manner that the source and purpose of the drawing can be readily identified;
■ contain all necessary routine information in a form that can be readily checked; and
■ be kept in comprehensive sets and stored easily. Record sets for drawings and documents issued for construction may have to be kept in archives for up to 15 years to take account of the long-stop period under the Latent Damage Act 1986. Note that this period is longer than the six year retention required by a contract executed under hand and 12 years required by a contract executed as a deed. Drawings and data may be stored digitally but care should be taken to

store such data in a generic format that future software will be able to access and on durable media.

An assortment of drawings of different shapes and sizes with title panels and essential information in differing positions is a cause of confusion and irritation. Standardisation should apply no matter how the drawings are prepared.

Size

Despite the fact that we create and develop designs on computer screens, there still remains the need to create paper-based images of drawings to communicate design intent to others and to create record copies with legal status.

In BS EN ISO 5457:1999 recommendations are made for drawing sheet sizes A0, A1, A2, A3 and A4. In selecting the standard drawing sizes, account should be taken of the many and varied methods of reproduction, for example, dye-line printing, photocopying, laser printing and photographic. Excluding routine dye-line printing, all methods available allow the possibility of changing the scale of the drawing in the course of reproduction – an invaluable asset in planning and producing comprehensive and co-ordinated sets of drawings. With this advantage in mind it is probable that the range of standard sheets might reasonably be limited to A0, A1, A3 and A4.

A0 is a rather unwieldy size, especially for handling on site, but it may be necessary for general arrangement drawings of large projects, townscape plans, extensive landscape drawings and full-size details. A1 size drawings are the most common and the most acceptable for ease of handling and can be reduced to any size down to A4. This is extremely useful in the preparation of design and presentation drawings that may be required in bound folders of A3 or A4 size. A3 sheets are most commonly used for quick sketches and details that may be reduced down to A4 size for use in transmission by fax.

With the almost universal use in offices of computers, the use of drawing sheets for the preparation of schedules is the exception rather than the rule. Standard letters, forms for instructions, certificates and most schedules are stored electronically for completion or amendment as the occasion arises. Drawing sheets might be used for window, door and ironmongery schedules that contain a considerable amount of information – in this case, A3 sheets would be the most convenient. The vast majority of drawings for a number of projects are produced in A3 size.

Layout and revision

The design of drawings must take account of both immediate and long-term storage. Whether it is for reduction and storage in files, in plan chests or in large clips hung on racks beside working stations, the layout of the sheet must provide sufficient margin to ensure that no information is obscured by the storage system. Many offices store drawings on servers accessible via local area networks and via the internet, dramatically reducing the amount of paper media stored and transported expensively. However, the need to have drawings printed on paper is still with us.

Part III

Two types of title and information panel are recommended in BS EN ISO 4157:1999. The examples include space for classification references, but this will not be required unless a classified form of specification is being used. The latest and most widely adopted classification is the Unified Classification for the Construction Industry (Uniclass). This is a comprehensive coded system providing an interface between the manufacturer's product database, the use of materials and methods of assembly shown on the drawings, the specification and the bills of quantities.

Seldom are drawings produced and left unchanged. The development of the design and collaboration between members of the design team both lead to a continuing process of amendment and revision and sustaining the effective collaboration between design team, contractor, facility manager and employer is an essential part of the BIM technique. Every revision to a drawing must be adequately described and the date when it was made noted on the sheet and indicated by a unique revision code. If this is not done, it will be difficult, if not impossible, for other members of the project team to discover changes made to the drawings. Despite the widespread use of computers with cheap storage, the use of broadband for fast transmission of large amounts of data and the use of extranets, it is still necessary to store project documentation in paper form for both practical and legal reasons. This will include the long term storage of 'as built' information.

Scale

Most printing processes distort drawings – some are reduced in size arbitrarily for the convenience of storage or transmission by fax and some may be microfilmed for long-term storage of 'as built' information. It is therefore good practice to incorporate clearly the appropriate drawn scale(s) on every original drawing. In any event it must be remembered that it is always dangerous to take dimensions from any drawing by scaling.

Difficulties can arise if drawings are prepared to unusual scales. Engineering consultants frequently use the scales of 1 : 25 and 1 : 250. This can cause considerable confusion and should be discouraged. It is recommended that the following scales be used:

Location plans

- 1 : 2500
- 1 : 1250
- 1 : 500

Site and development plans

- 1 : 500
- 1 : 200

General arrangement drawings

- 1 : 100
- 1 : 50

Part III

Component drawings

- 1 : 50
- 1 : 20

Details

- 1 : 5
- 1 : 1 (full size)

Assembly drawings

- 1 : 20
- 1 : 10

Nature and sequence of drawing production

The type of contract and the method of choosing the contractor can affect the nature of the drawings and the sequence of their production. The building consists of many elements – structural frame, walls, partitions, roof, mechanical services and so on. Each of these elements forms a part of the whole and only a complete set of drawings can inform the contractor as to the complete building. However, that does not necessarily mean that all drawings need to be completed before the contractor starts work on site – hence the need at the outset to understand the type of contract to be used.

The production of the drawings should be carefully planned. A preliminary list should be prepared of the drawings that will be required, not only from the architect but also from the other consultants.

RIBA Plan of Work 2013

The old RIBA Plan of Work was developed in 1963 and despite a number of updates in its 50 year history, it now fails to reflect current needs and practice to the extent that a significant new version is required. The Plan of Work 2013 provides such an update.

The RIBA in collaboration with CIC has developed a radically revised Plan of Work that will map BIM processes and at a more mundane level will alter the content of some drawings to respond appropriately to this new Plan. For example, Stage 3 (Developed Design) will produce the co-ordinated product of architects and all involved engineering disciplines. Stage 5 (Technical Design) takes account of the need for design by specialist sub-contractors and of performance-specified work.

Drawings for SBC contracts

When tenders are to be obtained and a contract placed on the basis of the SBC (or indeed JCT 98 which is still in fairly common use) with or without bills of quantities, all the drawings, schedules and specifications need to be completed prior to obtaining tenders, except when the contractor is provided with an information release schedule which states what information the architect will release and the time of its release. Reference to the RIBA Plan of Work (see Figure 2.1) will show how important it is to plan the production of the drawings, particularly through work stages E and F. It should also be borne in mind that where bills of quantities are used the NRM2 general rules set out details of the drawings required for the purposes of tendering.

The drawing production sequence is as follows:

- **General arrangement drawings:** Plans and elevations (1 : 100 or 1 : 50) which, after receiving the employer's approval, are sent to all consultants who will then prepare their draft schemes.
- **Services drawings by the architect (plumbing, drainage):** These are worked out in detail concurrently with the preparation of the general arrangement drawings.
- **Construction details by the architect and specialist sub-contractors:** The selection of materials and finishes made during the preparation of the general arrangement drawings leads to the preparation of the 'architect's details' and to obtaining tenders for specialist sub-contract design and construction. The outcome of this stage of the work can affect other aspects, particularly critical dimensions in the primary elements of the structure, and consequently the work of other consultants.
- **Assembly details:** When the drawings of various consultants are accepted, the assembly details (1 : 20 and larger if necessary), which have been drafted in outline, should be completed.
- **Final co-ordination of drawings:** When all the detailed information has been assembled and co-ordinated, the final overall picture (originally the outline general arrangement drawings) can be completed with accuracy.
- **Layout and site plans:** These are finally completed, incorporating information on all external services and the setting out of the buildings.

Many details and drawings will be changed during the course of the works and, in the process of adjusting the costs, it will be necessary to refer to the original information upon which the tender and contract sum were based. It is therefore of the utmost importance to keep a record set of the drawings used in the production of the other contract documents.

Drawings for design and build or management contracts

There are occasions when it is appropriate for the contractor to be responsible for the production of the building design or to manage a contract in such a manner as will accelerate the rate of construction (even though it might increase the cost). In such circumstances, rather than the preparation of the drawings being geared to the process of obtaining tenders for the whole of the works to be carried out by a single contractor, such drawing preparation will instead be geared towards obtaining tenders from sub-contractors for the various parts of the building, known as 'packages'.

Once the building is designed, the sub-contract packages involved will be identified. The general arrangement drawings will have the same significance as in the traditional procurement route but, once they are agreed, subsequent drawings will be of an elemental nature, that is, concerned with the various and separate elements of the building. The tenders for these packages will be obtained on the basis of drawings and specifications, and sometimes bills of quantities, for each package in a sequence related to the programme of work to be carried out on site. The order in which the drawings are produced will be related to the order of procurement and construction. Pile or foundation drawings, together with the main structural frame and drainage drawings, will be completely detailed and the work started on site, possibly before detailed or assembly drawings have started for secondary elements of the superstructure. Needless to say, the co-ordination of details in these circumstances becomes very difficult and can result in a rather different approach to design, calling for greater flexibility in the use of the structure or the building envelope to allow more readily for substantial changes to the design late in its development.

The following is an example of the order in which drawings, specifications and bills of quantities (when used) for packages might be required. Those items marked with an asterisk (*) typically have a long lead-in or delivery period:

- general arrangement drawings
- levelling and general site works
- structural frame (steelwork)*
- drainage
- foundations (including piling)
- lifts and escalators*
- proprietary walling*
- roofing systems
- windows and doors*
- brickwork details
- mechanical and electrical engineering services*
- plant and machinery details
- partitions
- staircases and other secondary elements
- ceilings
- fixtures and fittings
- external works.

There can be many more packages than are given here and some that are given separately may be combined.

Design intent information

Some of the newer forms of contract have their own requirements for the production of project documentation. For example, the JCT MP contract imposes demanding design submission procedures to amplify the design undertaken by the contractor.

In certain circumstances, the design team may be asked to produce design intent drawings. One example is that of a contractor working on a Private Finance Initiative (PFI) project as part of a consortium requiring information to pass to the in-house design team.

The paradox of design intent drawings is that in many instances they are virtually complete as they have to communicate specific designs, details and specifications to another professional team and will be used for value management and detailed costing purposes. They will also be reviewed by the end user's team to assess value for money in the PFI context.

It is important that the design team led by the architect determines, at the outset of the production information stage, the exact level of detail required to be given in the design intent set.

Computer aided design

The use of computers in architecture is now almost universal. The title given to this technique describes well the role of computers in the design process. Computers aid but they are not creative, rather they power-assist the creative process by enabling designers to assess their work accurately and rapidly. Computers are essential tools with many attractions to all the members of a design team. The main advantage lies in their ability to perform at great speed the technical services of producing drawings, of calculating and co-ordinating and ultimately of communicating. If there is a disadvantage, it may be seen in the type of drawings most commonly produced – of single line weight without enhancement or emphasis and often, for all their precision, lacking the special quality that allows easy reading and interpretation.

The most simple computer system will produce two-dimensional drawings containing no more information than is input for each drawing. The most advanced will store a complete model of the building and, on command, produce drawings to any scale or projection of any component part, section or the whole of the building. Moreover, amendments to the original model will automatically appear on all drawings as they are produced. The storage capacity of even modest computers can carry a huge library of standard components, assemblies and details which will enable the architect or technician quickly to produce a vast range of working drawings of great accuracy.

Where members of the design team are using compatible hardware, discs can be exchanged, giving each designer immediate access to all the information produced by the others. Layer upon layer, information can be added to the general arrangement drawings, ensuring complete co-ordination of all the parts of the building – structure, envelope details, mechanical and electrical engineering services and finishing details.

At the outset of the design process, members of the design team should discuss and agree the management of their CAD/BIM resources. All members of the team should produce one or more back-up copies of all drawings at all stages of production. Records of drawings undergoing major amendment should be stored.

Common systems for coding drawings might well be adopted. On large projects, where many thousands of drawings may be produced, it is possible for the designer to have his system linked to specialist agencies which are made responsible for cataloguing and distributing to members of the project team the many drawings as they are produced or revised.

CAD/BIM enables every conceivable calculation related to the design process to be carried out, including structural, thermal performance, energy management, fire safety and lighting. And by following the procedures of CPI (Co-ordinated Project Information), the designers' programmes can be interfaced with those of the quantity surveyor so that coded information on the drawings can be used in the production of the bills of quantities, specification documents and ongoing cost estimates.

The use of CAD/BIM does not change the ground rules for good practice, but the systematic approach that it requires does help to improve the clarity of the information produced and makes it more accessible. Whichever way the drawings are produced, the information they convey should be the same.

Drawing file formats and translation

A wide range of CAD/BIM systems, with varying software and training costs, is now in use, responding to the needs of architecture, interior architecture and the various engineering disciplines. With this diversity of software and capability come associated problems of the quality of the interface between disparate systems. The .dxf format is one method of storing information for use across a wide range of software platforms. One simple and widely occurring example is exchanging data between Microstation and AutoCAD/BIM formats. However, this method is not infallible and certain levels of detail may be lost in the translation process.

At the job set-up stage of any project, the preferred CAD/BIM platform should be agreed, and if commonality cannot be achieved, a translation protocol should be devised. Design files should be checked carefully after translation to avoid errors of omission or distortion.

Project extranets

One of the more significant problems in the construction industry is that of the quality of communication between members of the project team. The traditional system is inherently inefficient. In that setting, the employer was not part of the communication system and relied on maintaining a very close working relationship with the design team members to keep up with their work and the method of transmission of information between members of the project team had not really changed in centuries.

Project extranets have been developed to overcome the wide range of problems related to the access of members of the project team to the most up-to-date information. There are at least 170 extranet companies without commonality of format, although this volatile marketplace should, through the process of failures and amalgamations, settle down over time to serve the industry better.

Part III

Some systems are stronger during the design process while others perform best during construction. A project extranet is established along the lines of an intranet, using the Internet to form an electronic project community. A file server holds all project information produced by all participants. This information can be accessed by members of the project team (including the employer) on an on-demand basis from any location. Some extranets provide facilities for WebCams, so that site progress can be remotely checked by everyone concerned on a real-time basis and from anywhere in the world. Project information is classified according to stage, status and relevance. Access and the ability to manipulate information are carefully controlled. Users do not need to install a plethora of application software on their machines to view output from many sources. However, an important factor to consider when choosing a project extranet is to ensure that the chosen system will accommodate all the platforms used by the members of the project team. Most systems claim to run on browser technology but some operate only on Windows, thus excluding users of Unix, Macintosh and Linux. The need for a common 'engine' for all project extranets is important.

Above all, a project extranet changes fundamentally the relationship between the members of the project team. From members being passive recipients of information from one another, they become active participants, having to go to the extranet to extract information. This fundamental shift of responsibilities will improve the performance of the project team and ultimately the quality of service given to the employer.

Project extranets provide the members of the project team with the ability to

- share and organise project documents, plans and files but also set user-definable access control at the project, folder and document levels;
- access project information on a 24/7 basis in a secure, accessible, on-line location;
- automate the creation and distribution of project updates and document revisions;
- easily communicate with each other anywhere in the world;
- monitor project status and progress;
- access and view project audit trails and document version history;
- red-line documents, issue, control and manage requests for information and add comments to or indicate changes on project drawings, contracts and other documents;
- view on-line CAD/BIM output, photographs, spreadsheets, faxes and documents whilst maintaining data integrity;
- automatically maintain relationships within compound documents;
- access a complete document history, including copies of earlier document versions, revisions and access records;
- easily find specific documents within very large projects, via full text search or by document attributes such as file, name, type and author;
- access disaster recovery facilities for project data; and
- access project-cloning capabilities, enabling all projects to be organised in the same way and allowing the rapid implementation of new projects.

The use of extranets allows business process management techniques to be deployed in the design and construction process, with the advantage that the process

■ shortens project schedules;
■ reduces costs;
■ increases productivity;
■ extends best practice tools and techniques throughout a project;
■ reduces project risks; and
■ establishes full project accountability.

Some JCT contracts allow electronic communication, and the parties to contracts can agree to use such paradigms by entering into an Electronic Data Interchange (EDI) Agreement. Two standard forms exist: the EDI Association Standard EDI Agreement and the European Model EDI Agreement.

   Proprietary extranet software that are a core component of BIM software provide for structured permissions to read or alter design data. There are ever-present issues of the ownership, use and transfer of instruments of professional service, and the danger that contractors or employers regard design information as a 'product' with liability exposure. The reuse of data, the incorporation of design data into facility management programmes and the general dangers inherent in the electronic transfer of design data are ever present. The delivery of a drawing must not carry with it an express warranty or guarantee of exactness of information. Hard copies of drawings and data, archived by the architect, will retain control over amendments.

BIM Case Studies: Ministry of Justice

The Ministry of Justice has mandated the use of BIM on key capital projects and this technique is currently being used for the provision of prisons and law courts. Projects include HMP Cookham Wood, HMP Mount, HMP Aylesbury, Liverpool Magistrates Court and Aberystwyth Law Courts (although this project was stopped.) At HMP Cookham Wood the MoJ, in collaboration with its contractor partner Interserve, claimed to have saved £800,000 through the use of BIM. For the MoJ the benefits of BIM/extranet include:

■ achieving significant cost savings that will be increased even further as the use of BIM becomes more familiar and widespread;
■ improving collaboration between designers, constructors, facility managers and the employer for better outcomes;
■ efficiently and effectively sharing the correct information;
■ reducing errors and wastage;
■ facilitating the re-use of information;
■ reducing life cycle costs; and
■ improving the capture of knowledge.

> BIM Case Studies: Beacon South Quarter, Sandyford Co., Dublin
>
> BIM/extranet techniques were used to design buildings B2 (residential) and B3 (offices) by architects Traynor O'Toole for developer Landmark Developments with a combined construction cost of €55m. Close collaboration between the design team, employer, contractor and sub-contractors was via a hosted website with secure access. The software used was Autodesk 'Revit'. The architect reported increased team productivity, enhanced 3D visualisation and effective synchronisation between disciplines and constructors.

Contents of drawings

When planning and programming the production of drawings, account must be taken of the fact that every part of the building has to be designed, detailed and drawn. Obvious as this might seem, it is unfortunately only too common to find elegant sets of drawings repetitiously showing solutions to simple matters but assiduously avoiding the complicated. Should a part of a building be difficult to detail and to draw, that is the very detail and drawing that the contractor will need most! In considering how all the necessary information is to be conveyed, there are good and simple rules to observe:

- Draw every part of the building and do not repeat information unnecessarily.
- Sections are invaluable – indicate and code their location clearly.
- Remember that plans are merely horizontal sections; one plan for each floor level may not be enough.
- Number all rooms and spaces.
- Give reference numbers to all doors, windows, built-in fittings and the like; show these on all plans, sections and elevations.
- Make clear cross-reference to other drawings or to schedules, so that detailed information can be traced.
- Be consistent in showing structural grids and levels on all plans, sections and elevations.
- Provide concise notes where the text does not overlap with drawn information.
- Work out and indicate all essential dimensions but do not labour the obvious.

There follows a checklist of drawings and their contents.

Survey plan

- Existing site, boundaries and surrounds
- Positions of major features such as existing buildings, roads, streams, ponds, walls and gates
- Positions, girth, spread and type of trees
- Location and type of hedgerows
- Sufficient spot levels and contour lines, related to a specific datum, to enable a section of the site to be drawn in any direction
- Position, invert and cover or surface levels of existing drains

- Location, direction and depth of service mains
- Rights of way, bridle paths and public footpaths
- Access to site for vehicles
- Ordnance reference if available
- North point
- A key plan showing the relationship of the various sheets, should the survey cover more than one sheet.

Site plan, layout and drainage

- Relevant information from the survey plan
- Building profile and grid dimensionally related to a datum point or line
- New roads and paths with widths and levels marked
- Floor levels clearly indicated using the same datum as for the existing levels on the survey plan
- Steps and ramps where they occur
- Trees and hedgerows removed or retained
- Street and other external lighting
- Soil and surface water drains complete with pipe sizes and connections to sewers, making clear the distinction between soil and surface water drains and manholes (manhole sizes, levels and invert levels should be shown on a separate schedule)
- Gas, water and electric mains with depths indicated where possible and showing positions of connections to existing mains, supply company meters (external) and details of meter housings if required and points of termination within buildings
- Banking and cutting, and areas of disposing or spreading surplus soil
- Details of fencing, existing and new.

General arrangement

Once the design is approved and working drawings commenced, copy negatives should be taken of the general arrangement drawings of all floors showing only door swings in addition to walls and partitions and without any dimensions or other information on them. These drawings can then be developed without confusion or loss of time to show different details.

Services

- Incoming mains and meter positions for all services
- Electrical layout, including power points, light points, switches and all electrical equipment
- Heating layout showing boilers, calorifiers, radiators or similar heating equipment
- Plumbing and internal drainage layout
- Gas layout

Part III

- Sprinkler system layout, including pumps and storage tanks
- Fire detection and protection equipment
- Location of any special services.

Foundations

- The main building grid annotated and dimensioned
- Width and depth of all foundations for walls, piers and stanchions, with levels to underside
- Positions and levels of drains, gullies and manholes close to foundations
- Walls above foundations dotted with wall thicknesses dimensioned
- Positions of incoming service mains, service main ducts and trenches and their levels.

Floor plans

Complete plans drawn at a constant level through all openings for all floors and mezzanine floors with:

- Overall dimensions
- External dimensions, taking in all openings
- Internal dimensions to show the positions and thicknesses of internal walls, partitions and all major features
- Doors, their direction of swing and reference numbers
- Windows and their reference numbers
- Names or numbers of all rooms and circulation spaces (see BS 1192)
- Vertical and horizontal service ducts, holes in floor slabs, flues and builder's work in connection with services
- Staircases, their direction of rise and stair treads numbered
- Hatching to indicate materials of which walls and partitions are constructed
- Rainwater pipes
- Profile of joinery fittings and their code or reference numbers
- Sanitary fittings
- Floor levels clearly marked.

Roof plan

- Main construction features
- Levels clearly marked
- Types of covering
- Direction of falls
- Rainwater outlets, gutters and pipes
- Rooflights
- Tank rooms, trap doors, chimneys, ventilation pipes and other penetrations, builder's work in connection with services
- Parapets, copings and balustrades
- Duckboards, catwalks, escape stairs and ladders
- Lightning conductors, flag poles and aerials.

Elevations of all parts of the building

- New and old ground levels
- Profiles of foundations
- Damp-proof course levels
- Levels of ground and upper floor slabs
- External doors and windows with their reference numbers
- Air bricks and ventilators
- External materials, flashings
- Rainwater pipes and gutters
- Lightning conductors.

Descriptive sections

It is important to have sections that describe the juxtaposition of all elements of the building, adjacent buildings and the general site area. The sections should illustrate

- New and original ground levels showing cut and fill
- The main structural profile with grid lines
- Type and depths of foundations
- Floor and roof structures
- Windows, doors and rooflights with their reference numbers
- Lintels, arches and secondary structural members
- Damp-proof membranes and courses, flashings
- Eaves and valley gutters, parapets, rainwater heads and downpipes
- All vertical dimensions.

Ceiling plans at all floor levels

- Room names and numbers over which ceilings occur
- Type of ceiling, type of suspension, suspension pattern and finishes
- Height of ceiling above floor level
- Location of light fittings, their type and size
- Fire detection and safety installations, ceiling void compartmentation, smoke detectors, emergency lighting
- Mechanical ventilation registers
- Sprinkler layouts.

Construction details (scale 1 : 20 and 1 : 10)

- Detailed sections for external walls, foundations and roofs
- Plans, sections and elevations for staircases
- Lift wells and escalators

Part III

- Any room or part of the building, the setting out of which is difficult, involves extensive fittings, fixtures, plumbing or special features or requires careful co-ordination, such as
  - kitchens
  - bathrooms
  - lavatories
- Special-purpose rooms in hospitals, laboratories, etc.
- Window and door details
- Part elevations and sections of the building's envelope containing special features such as
  - entrances
  - special brick details
  - balconies
  - stonework
  - plant rooms
- Builder's work in connection with services
- Vertical and horizontal pipe ducts and their access
- Fireplaces and flues
- External details such as special paving, steps, kerbs, handrails and external lighting.

Large-scale details (scale 1 : 10 and 1 : 15)

These comprise enlargements of component parts of assemblies which have been shown at 1 : 20 scale but which require a larger scale to show the full details:

- Sills, heads and jambs of windows and doors
- Special brickwork, string courses, arches (rubbed bricks) and special flashing details
- Eaves, parapets, copings and special mouldings
- Timber sections such as handrails, window sections and joinery details
- Jointing details of curtain walling and other specialised cladding
- Staircases
- Special joinery fixtures and fittings
- Any special feature which cannot be described or shown clearly to a smaller scale.

Schedules

It is good practice to convey information on items such as windows, doors and ironmongery by means of schedules which will set out the architect's requirements for other members of the project team. This can be done in a manner which has many advantages over drawings, for example:

- Checking for errors of omission or duplication is simplified.
- Quantifying items for the purpose of obtaining estimates or placing orders is simplified.

- If the information is set down systematically, prolonged searching through a specification is avoided.

In setting down information in schedules, ensure clear layout, simple coding, the minimum of unfamiliar abbreviation and proper reference to the location in the building of the items concerned. Schedules should be either the same size as the standard drawing sheets selected for the job or, if smaller, bound together as a set in one of the accepted paper sizes.

The design of and information contained in the title blocks should be consistent with those of the drawings. Schedules should also be numbered in a manner which is an extension of the numbering system used for the drawings. The first schedule should be a list of all the drawings and schedules related to the project, with space to record the latest revision number of each document. The first formal use of this schedule will be its inclusion in the specification or bills of quantities providing a record of the drawings and schedules used in the preparation of the tender documents. Schedules are not intended to supplant the specification or bills of quantities, which will contain the full and final description of the characteristics of the items concerned.

The layout of schedules will vary according to the information to be conveyed. Items being scheduled may be listed to read downwards on the left-hand side, with the characteristics and sizes of the items reading across or vice versa. Diagrams may well be included in the body of the schedule and a column reserved for notes and records of revisions. Once an item has been fully described in a schedule, it should not be repeated at length. For example, having described the construction of a particular type of door, it can be given a type reference and only this would be repeated in the schedule. Characteristics, particularly on a finishes schedule, can often be defined by a code with an interpretation in the form of a key to one side. Constant repetition of descriptions is then eliminated. Well-recognised abbreviations which will not be confused with others can be used to save time.

Examples of schedules are given at the end of this chapter:

- Example 19.1 – Window schedule
- Example 19.2 – Door schedule
- Example 19.3 – Finishings schedule
- Example 19.4 – Manhole schedule.

This is by no means a complete and comprehensive list – there are many other items which may be described and collated conveniently in schedules, but it is very difficult for them to be standardised as they vary so much with each building. The examples given vary from those shown in BS EN ISO 4157:1999 but they are more easily understood and less likely to lead to error.

Drawings and schedules for records

Although we have already referred in this chapter to the keeping of record drawings, it is a matter of sufficient importance to justify emphasis and clarification. Before starting work on site there are two stages at which it is important to

Part III

WINDOWS

WINDOW TYPE REF:	A	B	C	D	E
WINDOW REF NO.	W2 to W19 inc. W22 to W39 inc.	W1, W20 W21, W40	W41, W42 and WW3	W44 to W50 inc.	W51
NUMBER OFF	36	4	3	7	1
MATERIAL OF MAIN FRAME	Extruded aluminium PVF2 treated, insulated sections: by Glazing International Limited				
SUB-FRAME	Nil	Nil	Nil	Nil	Hardwood
BUILDERS OPENING SIZE	Continuous horizontal opening height 1200		600 H × 1500 W	1800 H × 750 W	1800 diam
WINDOW UNIT SIZE	1185 H × 1450 W (inc. cill)	1185 H × 750 W (inc. cill)	585 H × 1485 W	1785 H × 735 W	1485 diam
FIXING JAMB	To coupling mullion	Alum. brackets to brickwork	Aluminium brackets to brickwork		
HEAD	Alumium brackets plugged to concrete lintals or bean				Nil
CILL	Aluminium cill with brackets plugged and screwed to blockwork				Nil
SEALANT	Brown polysulphide to heads and jambs				
IRONMONGERY AND GLAZING					
FASTENINGS	Alum. lever handle and locking latch	Nil	Aluminium level handle	Aluminium peg stay	Nil
OPENING LIGHTS HINGES/PIVOTS	Horizontal hung projecting opening gear	Nil	Top hung (hinges with frame)	Side hung (hinges with frame)	Nil
STAYS	As part of opening gear	Nil	Pair of friction stays	Nil	Nil
OPERATING GEAR	Nil	Nil	Nil	Nil	Nil
SECURITY	Safety catch in jamb	Nil	Nil	Nil	Nil
GLAZING	Double glazed units, outer glass 6 mm Pickertons Antison Silver, 12 mm void, inner glass 6 mm clear float				
FIXING	Extruded neoprene gaskets – all windows pre glazed by manufacturer				
NOTES					See drawing No. Do2/64

REVISIONS			DIAGRAMS				
Ref	Wind	Date					
A	WA/21	28/2/92					

Project Title   SHOPS & OFFICES – NEW BRIDGE STREET – BORCHESTER

Architects: REED & SEYMORE   WINDOW SCHEDULE   NO. 456/9.1   REV. A

Example 19.1   Window schedule.

Part III

DOORS

	A		B	C	D
DOOR TYPE REF:	A		B	C	D
DOOR REF NO.	1	2 to 8	1 to 4	1 to 8	1 to 15
LOCATION AND NUMBER	Main entrance lobby — 1	Entrance hall and corridor 01 & 11 — 7	Main staircase — 4	Rooms 03, 04, 05 & 10 to 14 incl. — 8	Offices 06 to 010 & 115 to 120 — 15
TYPE AND DESCRIPTION	HW framed single swing double door: 6 mm toughened glass		Plyfaced solid core flush panel with GWPP glazed viewing panel		Plyfaced solid core flush panel
FIRE RATING	Nil	Nil	60/60	30/30	Nil
DOOR SIZE	2 @ $2040 \times 726 \times 46$ with 12 mm reboted styles		$2040 \times 826 \times 46$ and $2040 \times 426 \times 46$	$2040 \times 826 \times 46$	
FINISH	Polyurathene sealed		Colour preservalive treated and wax polished		
FRAME OVERALL SIZE (NOMINAL) AND SECTION	2375×1850, 2375×1500 ex 150×60 rebated 25 mm kicking rail ex 200×36		2375×1350 ex 150×50 rebated 25 mm	2100×900 ex 150×60 rebated 25 mm	2100×900 125×32 plus stops
MATERIAL/ FINISH	Hardwood sealed		Softwood primed and painted		
FAN LIGHT	6 mm PP		6 mm GWPP	Nil	Nil
SIDE LIGHT	6 mm toughened glass	Nil	6 mm GWPP	Nil	Nil
SWING					
IRONMONGERY	1 @ 726 726	7 @ 726 726	4 @ 826 726	6 2	7 8
HINGES		1% pair SS		1 pair rising butts	1 pair
LOCKS AND LATCHES	Mortice deadlock type Ref. type –	Mortice dead lock type –	Mortice dead lock type –	Mortice latch and dead lock type –	
HANDLES	1 pair 'D' pattern pull handles type –		Lever handles type –		
KICK PLATES	2 pairs satin aluminium 200 high		Nil	Nil	Nil
PUSH PLATES	1 pair satin aluminium 200 high		Nil	Nil	Nil
BOLTS	2 off SA flush (to leaf without lock)		2 off SA flush to small leaf	Nil	Nil
STOPS AND STAYS	Skirting mounted type –			Floor mounted door stop type –	
MISCELLANEOUS			Surface mounted overhead closers. Intumescent strip to 3 sides		

REVISIONS

Ref	Door	Date
A	DA/2	28/2/92

DIAGRAMS

Project Title   SHOPS & OFFICES – NEW BRIDGE STREET – BORCHESTER

Architects: REED & SEYMORE   DOOR SCHEDULE   NO. 456/13.2   REV. A

Example 19.2   Door schedule.

FINISHES

FLOOR LEVEL	Ground	Ground	Ground	First	First
ROOM NAME	Entrance lobby	Stair lobby	Enquiries	General office	Womens lavatory
ROOM NO.	G01	G02	G03	101	102

FINISHES

SCREED AND FLOOR FINISH	75 mm screed Carpel, Swatch ZD 10016			Raised floor system with carpet inlays trays	50 mm screed 150 × 150 ceramic floor tite
WALLS 1 2 3 4	Glazed Sirpite plaster finish to sand and cement base coat				Plaster Glazed wall tiles Glazed wall tiles Glazed wall tiles
SKIRTINGS	4 × 100 × 19 HW as detail 456/W/26			19 × 100 SW	Flush, coved ceramic tite
CEILING	Concealed grid suspension system; Class 1 tiles		Lay – in tile suspension system: Class 0/1		Plaster board, scrim and set
MISC AND NOTES	Standard brass mat well frame set into screed	Staircase see detail drawing 456/14.3	Reception counter see drawing 456/W/25		

DECORATIONS

WALL FINISHES	Vinyl wall fabric by Sampsons Ltd	1 mist coat plus 2 coats vinyl emulsion – malt			Prime plus 2 coats flat oil
COLOURS 1 2 3 4	Glazed Ref No. SAM 1234 Ref No. SAM 1234 Ref No. SAM 1234	BS 00A01 BS 00A01 BS 00A01 BS 00A01	BS 20C37 BS 20C33 BS 20C33 BS 20C33	BS 16C33 BS 16C33 BS 16C33 BS 16C33	BS 18E51 2, 3 & 4 tiled
CEILINGS	Nil	Nil	Nil	Nil	Flat oil white
FRAMES AND ARCHITRAVES	1 + 2 coat gloss oil brilliant white				H2 gloss oil BS 20C40
DOORS	Glazed alumin	HW veneer 2 coats satin polyeurathene seal			Plaslic laminate faced
WINDOWS	Aluminium frames self finished throughout				
CILLS	1 + 2 coats gloss oil brilliant white				Tiled as walls
SKIRTINGS	1 + 2 coats gloss oil brilliant white				1 + 2 gloss oil BS 20C40

REVISIONS Ref Rm No. Date A C01 1/3/92 B I02 16/3/92	MISC AND NOTES	Handrails – 2 coats glass polyeurathen seal	Reception counter – American oak french polished		Plaslic laminate cubicles and ducting, extent of wall tiling etc. see drawing 456/W/27

Project Title   SHOPS & OFFICES – NEW BRIDGE STREET – BORCHESTER

Architects: REED & SEYMORE   FINISHINGS SCHEDULE   NO. 456/14.1   REV. B

Example 19.3 Finishings schedule.

MANHOLES AND COVERS

MANHOLE REF:	FM1	FM2	FM3	FM4	FM5
INIERNAL SIZE	1125 × 825	900 diam	1050 diam	1050 diam	1350 reducing to 900 diam
CONSTRUCTION AND MAKE	225 mm brick to 150 conc base	Precost concrete chamber rings, slabs and seating rings set in concrete backlill. 150 mm insitu concrete base			
COVER SIZE	450 × 600	450 × 600			
COVER TYPE	Med duty double – sealed	Grade C Fig 7		Grade C Fig 5	
INVERT LEVEL	7.315	7.214	7.02	6.807	5.705
GROUND LEVEL	8.46	8.153	8.175	8.08	8.07
COVER LEVEL	8.46	8.382	8.175	8.382	8.382
MAIN CHANNEL SIZE	100 mm	100 mm	100 mm	100/150 mm	150 mm
TYPE STRAIGHT (S) CURVED (C) TAPPERED (T)	S	C	S	S T	S
BRANCH CHANNELS SIZE TO RH	2 × 100 mm × 90°		1 × 100 mm × 90° 1 × 100 mm × 135°	1 × 150 mm × 90°	
SIZE TO LH	1 × 100 mm × 135°	1 × 100 mm × 90°	1 × 100 mm × 135°	1 × 100 mm × 90° 1 × 100 mm × 135°	1 × 100 mm × 90°
INTERCEPTING TRAP	100 mm				
F.A.I.					100 mm
BACKDROP INLET					As detail drawing 071/C
STEP IRONS					C.I. into precast units
MISCELLANEOUS	OPC between head of brick and G.F stab		MH cover roised to paving level on brick upstand		

REVISIONS			DIAGRAMS				
Ref	MH No	Date					
A	FM4	28/2/92					
			Internal MH				

Project Title   SHOPS & OFFICES – NEW BRIDGE STREET – BORCHESTER

Architects: REED & SEYMORE   MANHOLES & COVER SCHEDULE   NO. 456/   REV. A

Part III

Example 19.4 Manhole schedule.

keep record drawings and schedules before they are subjected to any further amendment:

■ Documents upon which the specification or bills of quantities are based – a full set of completed drawings and schedules.
■ Contract drawings and schedules – those to which the contract documents refer and which are marked accordingly.

It is useful for these copies to be kept in such form as permits them to be reproduced should the need arise.

True and accurate 'as built' drawings, including the exact routes of all services, are vital in the future maintenance of the building and copies should be made and issued to the employer upon completion of the contract as part of the health and safety file. Where an accurate 'as built' survey would prove uneconomical or unfeasible, the use of drawings 'issued for construction' is acceptable. Their preparation depends upon regular correction of the drawings as the work proceeds and as amendments are issued. They may be in plastic negative form so that they are easily reproduced and read. Microfilm can be used to minimise on storage space, although nowadays it is more than likely that the drawings will be stored and made available electronically.

Notes

1. Author of Sketching in Lead Pencil for Architects and Others (1925).
2. Latham M. (1994) Constructing the Team. London. HMSO ISBN 978-0-11-752994-6.
3. Egan J. (1998) Rethinking Construction: Report of the Construction Task Force London. HMSO

Chapter 20
Specifications

Nick Schumann

The use of specifications

The types and uses of specifications continue to change as the construction industry alters its working practices and seeks faster, better and cheaper delivery methods whilst also continuing to push the boundaries of design. Specifications remain essential contract documents which confirm what the purchaser (employer) is buying from the seller (contractor) and how design elements that remain incomplete at the point of contract will be finalised with minimal commercial and time consequence. Specifications form part of a suite of documents used during the procurement process, including the drawings, main contract conditions, bills of quantities (schedule of rates), conditions of tender, planning consents and so on. The specification is therefore an essential document in securing value for money through the proper definition of scope, method and quality requirements in any construction contract.

In some parts of the world, specifications still comprise preambles to the bills of quantities, or brief written statements clarifying issues which could not be shown on the drawings but these instances are becoming less common mainly due to the advent of lump sum or guaranteed fixed price contracts which are not re-measurable upon completion in consequence of which bills of quantities have now morphed into schedules of rates used for monthly valuations and the pricing of variations. Preambles (the part of the bills of quantities which specified some products) have as a consequence largely disappeared and as a consequence the bills of quantities are no longer the main source of specification information. The specification has become a key component in the suite of contract documentation detailing not only scope, materials and workmanship requirements but also visual, procedural, design, testing, quality control and responsibility issues – the complete picture. The specification can be considered as a prime document in any

The Aqua Group Guide to Procurement, Tendering and Contract Administration, Second Edition.
Edited by Mark Hackett and Gary Statham.
© 2016 John Wiley & Sons, Ltd. Published 2016 by John Wiley & Sons, Ltd.

construction contract, second only in importance to the terms and conditions, for the simple reason that it records 'the buy'.

In today's complex array of procurement, design and contracting options, which seek to secure robust lump sums utilising the expertise of many consultants, the specification confirms precisely what one party agrees to provide to the other in respect of scope, quality and method to ensure that the employer is actually getting what he is paying for. An inadequate clause or one which conflicts with clauses in other contract documents can lead to conflict and potential claims for additional costs, so an essential goal for any specification is to recognise the potential for such issues and to deal with them in a clear and concise way such that future disputes are minimised. In the modern world there tends to be a rush to start work on site to complete the project at the earliest possible date which very often means that the design is incomplete at the time of tender or contract. At the same time employers desire to pass as much risk onto the design team and contractor as possible, which in the 'faster is better' environment needs to be recognised and dealt with. The old fashioned process of 'design it, buy it, then build it' method has to be replaced by either a 'partially design it, buy it, finish the design and build it' or 'design it as you buy and build it' method which requires the specification to be written such that it recognises the process and adapts to it. Different procurement processes therefore require different specifications because 'one suit' does not fit all.

Over the last 20 years, the construction industry has tried many ways of dealing with the whole issue of risk transfer, including traditional single stage bid to one main contractor at stage F design, two stage bidding process to a main contractor and construction management but today the most popular procurement route is design and build. The main benefit of design and build is the early transfer of financial risk to a single party, that is, the contractor. This has had a significant impact on many aspects of construction including the specification, which has to take into account many new factors so as to remain a helpful and useful document.

Faster design and construction while maintaining employer demands for high quality and cost control has led to an increased use of off-site fabrication where factory environments facilitate better quality control, economies of scale and less time on site, all of which, in turn, require detailed design input from specialist manufacturers, which again affect how the specifications are written. Traditional construction documentation is not always relevant to or understood by manufacturers and specifications therefore have to be adapted to suit. Modern buildings involve very much of a team effort but legal responsibilities demand clear confirmations as to who does what – the specification fulfills that requirement.

Despite these changes in emphasis, control and consultant roles, there remain three basic ways to specify a construction project, all of which need to be adapted to suit the selected procurement route, which are:

1. By prescription.
2. By performance.
3. By design intent and performance.

Specifying by prescription

Otherwise known as detailed materials and workmanship specifications, prescriptive specifications represent a complete design solution where the designer is able to stipulate exactly what products are to be used by the contractor and how they should be installed. Such specifications pre-suppose that the design team know exactly what they want, have worked out exactly what is required and why, are allowed by European legislation to name products/manufacturers and are prepared to warrant the design, taking full responsibility for the end product achieving the employer's expectations. Prescriptive specifications are most commonly written for smaller projects using well-tried and tested technology.

Specifying by performance

Performance specifications in their purest form only specify performance requirements and as such are used when the designer has no visual requirements and instead requires the contractor to select suitable materials and methods to meet the stated performance criteria only. A performance specification does not state product or visual requirements, only what has to be achieved, and as such employers do not necessarily know what they are getting until after contract award. The designer's requirement is only that the end product performs as required, meets current legislation and complies with National Standards/Codes of Practice. As a rule of thumb an architect would only use a performance specification for elements that will not be seen in the permanent works.

Specifying by description

This type of hybrid specification has evolved for use when procuring specialist elements that require the designer, contractor and manufacturer to work together in order to provide a technical solution that complies with the architect's design intent and performance criteria. The main function of a descriptive specification is to define scope, design intent, procedures for completing detailed design and quality control and to provide the contractor with a fair indication of the solutions that are acceptable, acknowledging that the design is not complete at the point of contract such that contractual claims are minimised. Contractors and manufacturers are required to use their specialist expertise to complete the detailed design (in consultation with the design team), manufacture and install the works, providing full warranties and guarantees.

The need to prepare a contract specification, which fully represents 'the buy', is critical when specifying by description so that claims and variations are to be avoided at a later date. This requires a process of evaluating tenders in order to agree upon the key elements that reflect the scope of works and the method and parameters to be adopted during the completion of the detailed design in a post contract environment. Descriptive specifications require that the design team

Part III

and the contractor work in harmony and adopt a 'let's find the best solution' approach.

Large complex buildings tend to require the descriptive approach while the more traditional design solutions lend themselves to prescriptive documents.

Specification writing

Having established the various types and uses of specifications, there are a number of ways in which they can be prepared. Generally speaking, the following five processes are necessary in the production of a good specification: (i) decide on format; (ii) collect information; (iii) input information; (iv) check and test; and (v) deliver.

Decide on format

When selecting a format, three options are available – uniclass, masterformat or bespoke.

Uniclass

The Unified Classification for the Construction Industry published by RIBA Publications is a classification scheme for organising library materials and for structuring product literature and project information. At the time of writing Uniclass is in the process of metamorphosis into Uniclass2 but this chapter reflects the original Uniclass system. Section J, which comprises work sections for buildings, is reproduced at the end of this chapter as Figure 20.1 with the permission of RIBA Enterprises Ltd. Uniclass can also be used by the architect for the classification of drawings (see Chapter 19) and by the quantity surveyor, via NRM2, in the preparation of bills of quantities (see Chapter 22). It therefore provides useful links between the drawings, the bills of quantities (the pricing document) and the specification (the scope and quality document).

Care is required when using section A because it mixes specification requirements for the permanent works with preliminaries requirements necessary to allow tenderers to price temporary, non-measurable and time-related items. It is important to remember that in the UK (unlike the USA) the specification is rarely the pricing document and bills of quantities no longer serve as the specification. Whilst the two should be cross-referenced for ease of use, they should not be intermixed, although the specification may be incorporated in the bills of quantities in order to give it contract document status.

Another point worth noting is that the Uniclass work sections for buildings do not reflect construction trades or work packages. This results in most specifications being made up of several sections, forming a rather bulky document. However, the proper use of sections A (General specification for work packages) and Z (Building fabric reference specification) helps to reduce repetition.

Work sections for buildings   J

JA Preliminaries/General conditions	JB Complete buildings/structures/units	JD Groundwork

JA1 The project generally
- JA10 Project particulars
- JA11 Documentation
- JA12 The site/Existing buildings
- JA13 Description of the work

JA2 The Contract
- JA20 The Contract/Sub-contract

JA3 Employer's requirements
- JA30 Tendering/Sub-letting/Supply
- JA31 Provision, content and use of documents
- JA32 Management of the Works
- JA33 Quality standards/control
- JA34 Security/Safety/Protection
- JA35 Specific limitations on method/sequence/timing/use of site
- JA36 Facilities/Temporary works/Services
- JA37 Operation/Maintenance of the finished building

JA4 Contractor's general cost items
- JA40 Management and staff
- JA41 Site accommodation
- JA42 Services and facilities
- JA43 Mechanical plant
- JA44 Temporary works

JA5 Work by others or subject to instruction
- JA50 Work/Materials by the employer
- JA51 Nominated sub-contractors
- JA52 Nominated suppliers
- JA53 Work by statutory authorities
- JA54 Provisional work
- JA55 Dayworks

JA6 Preliminaries for specialist contracts
- JA60 Demolition contract preliminaries
- JA61 Ground investigation contract preliminaries
- JA62 Piling contract preliminaries
- JA63 Landscape contract preliminaries

JA7 General specification for work packages
- JA70 General specification for building fabric work
- JA71 General specification for building services work

JB1 Prefabricated buildings/structures/units
- JB10 Prefabricated buildings/structures
- JB11 Prefabricated building units

JC Existing site/buildings/services

JC1 Investigations/Surveys
- JC10 Site survey
- JC11 Ground investigation
- JC12 Underground services survey
- JC13 Building fabric survey
- JC14 Building services survey

JC2 Demolition/Removal
- JC20 Demolition
- JC21 Toxic/hazardous material removal

JC3 Alteration – support
- JC30 Shoring/Facade retention

JC4 Repairing/Renovating/Conserving concrete/masonry
- JC40 Cleaning masonry/concrete
- JC41 Repairing/Renovating/Conserving masonry
- JC42 Repairing/Renovating/Conserving concrete
- JC45 Damp proof course renewal/insertion

JC5 Repairing/Renovating/Conserving metal/timber
- JC50 Repairing/Renovating/Conserving metal
- JC51 Repairing/Renovating/Conserving timber
- JC52 Fungus/Beetle eradication

JC9 Alteration – composite items
- JC90 Alterations – spot items

JD1 Ground stabilisation/dewatering
- JD11 Soil stabilisation
- JD12 Site dewatering

JD2 Excavation/filling
- JD20 Excavating and filling
- JD21 Landfill capping

JD3 Piling
- JD30 Piling

JD4 Ground retention
- JD40 Embedded retaining walls
- JD41 Crib walls/Gabions/Reinforced earth

JD5 Underpinning
- JD50 Underpinning

JE In situ concrete/Large precast concrete

JE0 Concrete construction generally
- JE05 In situ concrete construction generally

JE1 Mixing/Casting/Curing/Spraying in situ concrete
- JE10 Mixing/Casting/Curing in situ concrete
- JE11 Sprayed concrete

JE2 Formwork
- JE20 Formwork for in situ concrete

JE3 Reinforcement
- JE30 Reinforcement for in situ concrete
- JE31 Post tensioned reinforcement for in situ concrete

JE4 In situ concrete sundries
- JE40 Designed joints in in situ concrete
- JE41 Worked finishes/Cutting to in situ concrete
- JE42 Accessories cast into in situ concrete

JE5 Structural precast concrete
- JE50 Precast concrete frame structures

JE6 Composite construction
- JE60 Precast/Composite concrete decking

Part III

Figure 20.1   Uniclass section J (RIBA enterprise).

J   **Work sections for buildings**

JF	**Masonry**

JF1   **Brick/Block walling**
    JF10   Brick/Block walling
    JF11   Glass block walling

JF2   **Stone walling**
    JF20   Natural stone rubble
           walling
    JF21   Natural stone ashlar
           walling/dressings
    JF22   Cast stone walling/
           dressings

JF3   **Masonry accessories**
    JF30   Accessories/Sundry items
           for brick/block/stone
           walling
    JF31   Precast concrete sills/
           lintels/copings/features

JG	**Structural/Carcassing metal/ timber**

JG1   **Structural/Carcassing metal**
    JG10   Structural steel framing
    JG11   Structural aluminium
           framing
    JG12   Isolated structural metal
           members

JG2   **Structural/Carcassing timber**
    JG20   Carpentry/Timber
           framing/First fixing

JG3   **Metal/Timber decking**
    JG30   Metal profiled sheet
           decking
    JG31   Prefabricated timber unit
           decking
    JG32   Edge supported/
           Reinforced woodwool
           slab decking

JH	**Cladding/Covering**

JH1   **Glazed cladding/covering**
    JH10   Patent glazing
    JH11   Curtain walling
    JH12   Plastics glazed vaulting/
           walling
    JH13   Structural glass
           assemblies
    JH14   Concrete rooflights/
           pavement lights
    JH15   Rainscreen cladding/
           overcladding

JH2   **Sheet/board cladding**
    JH20   Rigid sheet cladding
    JH21   Timber weatherboarding

JH3   **Profiled/flat sheet cladding/ covering**
    JH30   Fibre cement profiled
           sheet cladding/covering
    JH31   Metal profiled/flat sheet
           cladding/covering
    JH32   Plastics profiled sheet
           cladding/covering
    JH33   Bitumen and fibre
           profiled sheet cladding/
           covering

JH4   **Panel cladding**
    JH40   Glass reinforced cement
           panel cladding/features
    JH41   Glass reinforced plastics
           panel cladding/features
    JH42   Precast concrete panel
           cladding/features
    JH43   Metal panel cladding/
           features

JH5   **Slab cladding**
    JH50   Precast concrete slab
           cladding/features
    JH51   Natural stone slab
           cladding/features
    JH52   Cast stone slab cladding/
           features

JH6   **Slate/Tile cladding/covering**
    JH60   Plain roof tiling
    JH61   Fibre cement slating
    JH62   Natural slating
    JH63   Reconstructed stone
           slating/tiling
    JH64   Timber shingling
    JH65   Single lap roof tiling
    JH66   Bituminous felt shingling

JH7   **Malleable sheet coverings/cladding**
    JH70   Malleable metal sheet
           pre-bonded coverings/
           claddings
    JH71   Lead sheet coverings/
           flashings
    JH72   Aluminium sheet
           coverings/flashings
    JH73   Copper strip/sheet
           coverings/flashings
    JH74   Zinc strip/sheet
           coverings/flashings
    JH75   Stainless steel strip/sheet
           coverings/flashings
    JH76   Fibre bitumen
           thermoplastic sheet
           coverings/flashings

JH9   **Other cladding/covering**
    JH90   Tensile fabric coverings
    JH91   Thatch roofing

JJ	**Waterproofing**

JJ1   **Cementitious coatings**
    JJ10   Specialist waterproof
           rendering

JJ2   **Asphalt coatings**
    JJ20   Mastic asphalt tanking/
           damp proofing
    JJ21   Mastic asphalt roofing/
           insulation/ finishes
    JJ22   Proprietary roof decking
           with asphalt finish

JJ3   **Liquid applied coatings**
    JJ30   Liquid applied tanking/
           damp proofing
    JJ31   Liquid applied waterproof
           roof coatings
    JJ32   Sprayed vapour barriers
    JJ33   In situ glass reinforced
           plastics

JJ4   **Felt/flexible sheets**
    JJ40   Flexible sheet
           tanking/damp proofing
    JJ41   Built-up felt roof
           coverings
    JJ42   Single layer polymeric
           roof coverings
    JJ43   Proprietary roof decking
           with felt finish
    JJ44   Sheet linings for pools/
           lakes/waterways

JK	**Linings/Sheathing/Dry partitioning**

JK1   **Rigid sheet sheathing/linings**
    JK10   Plasterboard dry lining/
           partitions/ceilings
           For panel partitions see
           JK30.
    JK11   Rigid sheet flooring/
           sheathing/linings/casings
    JK12   Under purlin/Inside rail
           panel linings
    JK13   Rigid sheet fine linings/
           panelling
    JK14   Glass reinforced gypsum
           linings/panelling/casings/
           mouldings
    JK15   Vitreous enamel linings/
           panelling

JK2   **Timber board/Strip linings**
    JK20   Timber board flooring/
           sheathing/linings/casings
    JK21   Timber strip/board fine
           flooring/linings

Figure 20.1   (*continued*)

Work sections for buildings   J

JK3   Dry partitions
JK30 Panel partitions
JK31 *intentionally not used*
JK32 Framed panel cubicles
JK33 Concrete/Terrazzo
      partitions

JK4   False ceilings/floors
JK40 Demountable suspended
      ceilings
JK41 Raised access floors

JL   Windows/Doors/Stairs

**JL1   Windows/Rooflights/
Screens/Louvres**
JL10 Windows
JL11 Rooflights/ Roof windows
JL12 Screens
JL13 Louvred ventilators
JL14 External louvres/
      shutters/canopies/blinds

JL2   Doors/Shutters/Hatches
JL20 Doors
JL21 Shutters
JL22 Hatches

JL3   Stairs/Walkways/Balustrades
JL30 Stairs/Walkways/
      Balustrades

JL4   Glazing
JL40 General glazing
JL41 Lead light glazing
JL42 Infill panels/sheets

JM   Surface finishes

JM1   Screeds/Trowelled flooring
JM10 Cement:sand/Concrete
      screeds/toppings
JM11 Mastic asphalt flooring/
      floor underlays
JM12 Trowelled bitumen/resin/
      rubber-latex flooring
JM13 Calcium sulfate based
      screeds

JM2   Plastered coatings
JM20 Plastered/Rendered/
      Roughcast coatings
JM21 Insulation with rendered
      finish
JM22 Sprayed monolithic
      coatings
JM23 Resin bound mineral
      coatings

**JM3   Work related to plastered
coatings**
JM30 Metal mesh lathing/
      Anchored reinforcement
      for plastered coatings
JM31 Fibrous plaster

JM4   Rigid tiles
JM40 Stone/Concrete/Quarry/
      Ceramic tiling/Mosaic
JM41 Terrazzo tiling/In situ
      terrazzo
JM42 Wood block/Composition
      block/Parquet flooring

JM5   Flexible sheet/tile coverings
JM50 Rubber/Plastics/Cork/
      Lino/Carpet tiling/
      sheeting
JM51 Edge fixed carpeting
JM52 Decorative papers/fabrics

JM6   Painting
JM60 Painting/Clear finishing
JM61 Intumescent coatings for
      fire protection of
      steelwork

JN   Furniture/Equipment

**JN1   General purpose fixtures/
furnishings/equipment**
JN10 General fixtures/
      furnishings/equipment
JN11 Domestic kitchen fittings
JN12 Catering equipment
JN13 Sanitary appliances/
      fittings
JN14 Plant containers
JN15 Signs/Notices

**JN2   Special purpose fixtures/
furnishings/equipment**
JN20 Appropriate section title
      for each project
JN21 Appropriate section title
      for each project
JN22 Appropriate section title
      for each project
JN23 Appropriate section title
      for each project

JP   Building fabric sundries

JP1   Sundry proofing/insulation
JP10 Sundry insulation/
      proofing work/fire stops
JP11 Foamed/Fibre/Bead cavity
      wall insulation

JP2   Sundry finishes/fittings
JP20 Unframed isolated
      trims/skirtings/sundry
      items
JP21 Ironmongery
JP22 Sealant joints

**JP3   Sundry work in connection with
engineering services**
JP30 Trenches/Pipeways/Pits
      for buried engineering
      services
JP31 Holes/Chases/Covers/
      Supports for services

**JQ   Paving/Planting/Fencing/Site
furniture**

**JQ1   Edgings/Accessories for
pavings**
JQ10 Kerbs/Edgings/Channels/
      Paving accessories

JQ2   Pavings
JQ20 Granular sub-bases to
      roads/pavings
JQ21 In situ concrete
      roads/pavings/bases
JQ22 Coated macadam/
      Asphalt roads/pavings
JQ23 Gravel/Hoggin/Bark
      roads/pavings
JQ24 Interlocking brick/block
      roads/pavings
JQ25 Slab/Brick/Sett/Cobble
      pavings
JQ26 Special surfacings/
      pavings for sport/general
      amenity

JQ3   Planting
JQ30 Seeding/Turfing
JQ31 Planting
JQ32 Planting in special
      environments
JQ35 Landscape maintenance

JQ4   Fencing
JQ40 Fencing

JQ5   Site furniture
JQ50 Site/Street furniture/
      equipment

Part III

Figure 20.1 *(continued)*

J   **Work sections for buildings**

JR	Disposal systems

JR1   Drainage
JR10   Rainwater pipework/gutters
JR11   Foul drainage above ground
JR12   Drainage below ground
JR13   Land drainage
JR14   Laboratory/Industrial waste drainage

JR2   Sewerage
JR20   Sewage pumping
JR21   Sewage treatment/sterilisation

JR3   Refuse disposal
JR30   Centralised vacuum cleaning
JR31   Refuse chutes
JR32   Compactors/Macerators
JR33   Incineration plant

JS	Piped supply systems

JS1   Water supply
JS10   Cold water
JS11   Hot water
JS12   Hot and cold water (small scale)
JS13   Pressurised water
JS14   Irrigation
JS15   Fountains/Water features

JS2   Treated on site water supply
JS20   Treated/Deionised/Distilled water
JS21   Swimming pool water treatment

JS3   Gas supply
JS30   Compressed air
JS31   Instrument air
JS32   Natural gas
JS33   Liquefied petroleum gas
JS34   Medical/Laboratory gas

JS4   Petrol/Oil storage
JS40   Petrol/Diesel storage/distribution
JS41   Fuel oil storage/distribution

JS5   Other supply systems
JS50   Vacuum
JS51   Steam

JS6   Fire fighting – water
JS60   Fire hose reels
JS61   Dry risers
JS62   Wet risers
JS63   Sprinklers
JS64   Deluge
JS65   Fire hydrants

JS7   Fire fighting – gas/foam
JS70   Gas fire fighting
JS71   Foam fire fighting

JT	Mechanical heating/Cooling/Refrigeration systems

JT1   Heat source
JT10   Gas/Oil fired boilers
JT11   Coal fired boilers
JT12   Electrode/Direct electric boilers
JT13   Packaged steam generators
JT14   Heat pumps
JT15   Solar collectors
JT16   Alternative fuel boilers

JT2   Primary heat distribution
JT20   Primary heat distribution

JT3   Heat distribution/utilisation – water
JT30   Medium temperature hot water heating
JT31   Low temperature hot water heating
JT32   Low temperature hot water heating (small scale)
JT33   Steam heating

JT4   Heat distribution/utilisation – air
JT40   Warm air heating
JT41   Warm air heating (small scale)
JT42   Local heating units

JT5   Heat recovery
JT50   Heat recovery

JT6   Central refrigeration/Distribution
JT60   Central refrigeration plant
JT61   Chilled water

JT7   Local cooling/Refrigeration
JT70   Local cooling units
JT71   Cold rooms
JT72   Ice pads

JU	Ventilation/Air conditioning systems

JU1   Ventilation/Fume extract
JU10   General ventilation
JU11   Toilet ventilation
JU12   Kitchen ventilation
JU13   Car parking ventilation
JU14   Smoke extract/Smoke control
JU15   Safety cabinet/Fume cupboard extract
JU16   Fume extract
JU17   Anaesthetic gas extract

JU2   Industrial extract
JU20   Dust collection

JU3   Air conditioning – all air
JU30   Low velocity air conditioning
JU31   VAV air conditioning
JU32   Dual-duct air conditioning
JU33   Multi-zone air conditioning

JU4   Air conditioning – air/water
JU40   Induction air conditioning
JU41   Fan-coil air conditioning
JU42   Terminal re-heat air conditioning
JU43   Terminal heat pump air conditioning

JU5   Air conditioning – hybrid
JU50   Hybrid system air conditioning

JU6   Air conditioning – local
JU60   Air conditioning units

JU7   Other air systems
JU70   Air curtains

JV	Electrical supply/power/lighting systems

JV1   Generation/Supply/HV distribution
JV10   Electricity generation plant
JV11   HV supply/distribution/public utility supply
JV12   LV supply/public utility supply

JV2   General LV distribution/lighting/power
JV20   LV distribution
JV21   General lighting
JV22   General LV power

JV3   Special types of supply/distribution
JV30   Extra low voltage supply
JV31   DC supply
JV32   Uninterrupted power supply

Figure 20.1   (*continued*)

Work sections for buildings   J

JV4   Special lighting
JV40   Emergency lighting
JV41   Street/Area/Flood lighting
JV42   Studio/Auditorium/Arena
        lighting

JV5   Electric heating
JV50   Electric underfloor/ceiling
        heating
JV51   Local electric heating
        units

JV9   General/Other electrical work
JV90   General lighting and
        power (small scale)

JX	Transport systems

JX1   People/Goods
JX10   Lifts
JX11   Escalators
JX12   Moving pavements
JX13   Powered stairlifts
JX14   Fire escape chutes/slings

JX2   Goods/Maintenance
JX20   Hoists
JX21   Cranes
JX22   Travelling cradles/
        Gantries/Ladders
JX23   Goods distribution/
        Mechanised warehousing

JX3   Documents
JX30   Mechanical document
        conveying
JX31   Pneumatic document
        conveying
JX32   Automatic document
        filing and retrieval

JW	Communications/Security/Control systems

JW1   Communications – speech/audio
JW10   Telecommunications
JW11   Paging/Emergency call
JW12   Public address/
        Conference audio
        facilities

JW2   Communications – audio-visual
JW20   Radio/TV/CCTV
JW21   Projection
JW22   Information/Advertising
        display
JW23   Clocks

JW3   Communications – data
JW30   Data transmission

JW4   Security
JW40   Access control
JW41   Security detection and
        alarm

JW5   Protection
JW50   Fire detection and alarm
JW51   Earthing and bonding
JW52   Lightning protection
JW53   Electromagnetic
        screening
JW54   Liquid detection alarm
JW55   Gas detection alarm
JW56   Electronic bird/vermin
        control

JW6   Central control
JW60   Central control/Building
        management

JY	Services reference specification

JY1   Pipelines and ancillaries
JY10   Pipelines
JY11   Pipeline ancillaries

JY2   General pipeline equipment
JY20   Pumps
JY21   Water tanks/cisterns
JY22   Heat exchangers
JY23   Storage cylinders/
        Calorifiers
JY24   Trace heating
JY25   Cleaning and chemical
        treatment

JY3   Air ductlines and ancillaries
JY30   Air ductlines/ancillaries

JY4   General air ductline equipment
JY40   Air handling units
JY41   Fans
JY42   Air filtration
JY43   Heating/Cooling coils
JY44   Air treatment
JY45   Silencers/Acoustic
        treatment
JY46   Grilles/Diffusers/Louvres

JY5   Other common mechanical items
JY50   Thermal insulation
JY51   Testing and
        commissioning of
        mechanical services
JY52   Vibration isolation
        mountings
JY53   Control components –
        mechanical
JY54   Identification –
        mechanical
JY59   Sundry common
        mechanical items

JY6   Cables and wiring
JY60   Conduit and cable
        trunking
JY61   HV/LV cables and wiring
JY62   Busbar trunking
JY63   Support components –
        cables

JY7   General electrical equipment
JY70   HV switchgear
JY71   LV switchgear and
        distribution boards
JY72   Contactors and starters
JY73   Luminaires and lamps
JY74   Accessories for electrical
        services

JY8   Other common electrical items
JY80   Earthing and bonding
        components
JY81   Testing and
        commissioning of
        electrical services
JY82   Identification – electrical
JY89   Sundry common
        electrical items

JY9   Other common mechanical and/or electrical items
JY90   Fixing to building fabric
JY91   Off-site painting/Anti-
        corrosion treatments
JY92   Motor drives – electric

JZ	Building fabric reference specification

JZ1   Fabricating
JZ10   Purpose made joinery
JZ11   Purpose made metalwork
JZ12   Preservative/Fire
        retardant treatments for
        timber

JZ2   Fixing/Jointing
JZ20   Fixings/Adhesives
JZ21   Mortars
JZ22   Sealants

JZ3   Finishing
JZ30   Off-site painting
JZ31   Powder coatings
JZ32   Liquid coatings
JZ33   Anodising

Figure 20.1   (*continued*)

Part III

Masterformat

The Masterformat classification system is produced by the Construction Specifications Institute in the USA. It is the system that is recognised worldwide and is best adopted when working outside the UK or when working with large American firms in the UK. Masterformat is more construction trade related than Uniclass but it does include Section 1 (General Condition), which is similar to preliminaries in the UK. It is worth noting that bills of quantities are not generally recognised in the USA and that the specification is instead the document used for pricing.

Bespoke

Major employers commissioning multiple buildings occasionally produce specifications tailored to suit their particular needs, based on building type/size or a partnering procurement strategy. In most situations, however, the bespoke approach is not recommended, as it is preferable to follow a recognised published format.

Collect information

In essence the specification is a risk management and information conveyancing tool that provides design and process requirements from one party to another for contractual and control purposes. Accurate information is essential in order to provide 'best information'. This can take many forms, ranging from the stipulation of a precise material and manufacturer to a detailed description of design intent. Both are equally effective provided that the specification format suits the procurement method and the selected form of contract. The wrong level of information put into a specification can lead to variations and claims for loss and/or expense.

One good method of collection is to produce a list of materials or system descriptions, each with a unique short code such as 'BLK-1 = 100 mm common blockwork'. These codes can then be used on the drawings and in the specification to provide clear scope and definition. This system is commonly known as technical sheet collection.

The design team should decide at an early stage how best to gather information, bearing in mind the time available. Start collection early. Information should be updated continuously until the specification issue date. Continuous drafting and updating in parallel with drawing production is the best methodology. All changes should be recorded for quality control purposes.

Input information

The key to accurate and fast completion of a specification is to use reliable and up-to-date source data. Off-the-shelf products are available. The best for use in the UK is perhaps the National Building Specification (NBS) which is arranged under the Uniclass work sections for buildings. It comprises detailed specification clauses together with skeleton clauses providing a sound checklist of the information necessary to specify properly preliminaries, general conditions, materials, goods and workmanship. NBS is updated regularly to take account of amendments to the JCT forms and to reflect changes in technology and the latest standards for materials, goods and workmanship.

Masterspec is perhaps the best product for use in the USA when adopting Masterformat. Web-based products are also becoming available, which usually provide up-to-date information and a fast production system. Whatever the source data, it is good practice for every clause in the specification to be uniquely numbered and for the project title, work section name and/or reference, version number, issue date and author's name to be given on every page. It is also good practice for drafts of the specification to be issued regularly for review and co-ordination by the design team.

Check and test

It is essential to ensure that there will be sufficient time available in which to test and check the specification before delivery. The author of a specification will never actually know how good it is until it is in use as a working document and is shown to provide clear direction when problems are encountered on site. However, it is a worthwhile exercise to test all specifications prior to finalising them to ensure that they include the appropriate clauses adequately to resolve all problems previously encountered in work of a similar nature.

As for checking, it is almost impossible for the same person to be able to write and accurately proofread a document because authors rarely spot their own mistakes. It is therefore essential to have a specification read through by a third party, who is preferably himself a competent specification writer and able to spot not only typographical errors but also where a mistake has been made or something does not make sense.

Deliver

There are two ways to deliver (issue) a specification – hard copy or electronic copy.

Hard copy

This involves numerous paper copies being delivered to all users and interested parties. These copies require close control and a record must be kept of exactly which version has been issued and to whom, when and for what purpose. This method of delivery is expensive, slow and cumbersome and uses vast quantities of paper.

Electronic copy

The construction industry is adapting to technology and the use of electronic document management systems and web-based sites as a means of recording, storing and circulating information is on the increase. The use of portable electronic memory devices to circulate documents gives rise to similar problems as the use of hard copy, except there is less bulk and less paper used. The use of email is preferable as it is fast, reliable and cheap. However, document identification and circulation control are just as important as with hard copy. Also, special consideration should be given to format – each recipient needs to be using the same software and settings in order to ensure that what everyone actually sees is the same document. It is strongly recommended that electronic files be issued always in a protected unexecutable format, such as PDF, in order to ensure that recipients cannot readily make inappropriate alterations – when that happens it can be disastrous. If any amendments to the specification are required for any reason, these should be raised with the design team and the specification ought only to be altered with express instruction.

Web-based, dynamic production systems are becoming available, providing faster, cheaper and better quality methods of production.

BIM

The adoption of BIM is set to change the way we share, exchange and use information and should considerably improve construction efficiency by driving out waste, reducing risk and saving time. BIM provides the opportunity for specification information to be embedded into design models along with costing information. Bringing the millions of generic descriptions, products and related information together for use in BIM will be a challenge and a number of products are already planned and built. Time will tell how successful this will be but it will change the ways in which specifications are prepared, delivered and used.

BIM will not, however, replace human input and talent, which will remain necessary to improve the built environment, nor will it change the need for expertise in preparing appropriate specifications to suit building types, different procurement routes, various market conditions, local capability and so on.

The reader's attention is drawn to Chapter 21 for a more detailed treatment of BIM.

Chapter 21
Building Information Modelling

Paul Morrell OBE

The BIM revolution – what is BIM, and who/what is it for?

Let's start with the name and all of the many things that Building information modelling (BIM) might mean.

The definition favoured by the UK Construction Project Information Committee is 'the digital representation of physical and functional characteristics of a facility, creating a shared knowledge resource for information about it, forming a reliable basis for decisions during its life cycle, from early conception to demolition'[1]; however, it is perhaps best described in summary as the intelligent use of information technology to design, construct and manage built assets.

Some people argue that it is mis-named because it isn't just about buildings. Indeed, it applies equally to all built assets and is currently critical to the delivery of some of our major infrastructure projects, such as Crossrail and the Thames Tideway Tunnel. Nor is it only about the building process, as its use stretches right through the project lifecycle. And finally, nor is it just about modelling, as it relates equally to the management of the process and of the completed asset.

What BIM *is* about is information (or data), and the production, management and distribution of that information so that everybody involved in the complex business of delivering a construction project can be fed with the information that they need to do the job and can feed in the information that they produce for everybody else who needs it. In fact, it is a useful frame of mind for clients and their advisers considering BIM to regard the procurement and provision of data as an integral part of the procurement of the asset itself.

As so many of the things that go wrong in projects relate to a breakdown in the flow of information, because what is needed is missing or incomplete or uncoordinated or not communicated to the right people at the right time, it also has to be accepted that fixing this problem goes beyond software, and it's worth saying at

The Aqua Group Guide to Procurement, Tendering and Contract Administration, Second Edition.
Edited by Mark Hackett and Gary Statham.
© 2016 John Wiley & Sons, Ltd. Published 2016 by John Wiley & Sons, Ltd.

the outset that the real potential of BIM will be neither appreciated nor captured unless it is understood that it is about a lot more than information technology or easing the production of design and construction information.

In the first instance, though, BIM should be regarded as the portal through which the building industry might also seek to make some of the progress with the application of ICT into its processes that have made such changes in other industries. Hitherto, every project has concerned enormous quantities of paper, band width and time in shuffling records between and within businesses as firms are prequalified, quotations are sought, orders are placed, deliveries are confirmed, time is recorded, applications for payment are submitted and processed – and so on almost ad infinitum. The potential for reducing both cost and error through, for example, electronic tendering and invoicing is therefore enormous, and although this is not specifically covered by most definitions of BIM, it is the reform that accompanies the adoption of BIM that should motivate greater and more innovative use of ICT.

To repeat, though, everything that we want from BIM cannot be bought in a box, and BIM alone will not bring about the change necessary to produce both a better industry and better buildings. That calls for leadership, changes in culture, organisation and processes that are deeply embedded in the traditional ways of doing business in construction and the accumulation of new habits that go far beyond the use of technology; that is a far greater challenge than trying to get some software to work – difficult enough though even that may be. The prize is, however, so great that it justifies almost any amount of effort to make it work.

This prize amounts to no less than an improvement in both the level and predictability of the cost, time and quality of a project – and therefore its value. This improvement will come principally through:

- all project team members working with the same data;
- the reduction in transaction costs and errors arising from the transfer or repeated input of data;
- the co-ordination of design information – for example, through clash detection programmes;
- supporting a move towards other desiderata in improved industry performance, such as greater standardisation and the use of off-site fabrication;
- the many checks and balances possible through the ability to test design proposals in a computer before starting on site, reducing the risks and costs of failure that can result from untested ideas; and
- the better communication of ideas, design proposals and the project as constructed via the three-dimensional model.

With something as complex as the creation of a new functioning asset, and with the particular risks that attend an undertaking in which virtually every project is a prototype, it simply makes no sense to try things for the first time at a scale of 1:1 in the wind and rain, when many of the characteristics of the design, and its construction and subsequent operation, can be tested in virtual reality first.

As three-dimensional models improve, and practitioners build on the possibilities of the technology, there will be an increase in dynamic uses of the model. These uses extend during both construction and occupation, and examples include the following:

- project visualisation, both for project owners and for wider consultation – such as public reaction to new developments;
- predicting building performance – such as structural integrity, fire resistance, temperature control, air quality, energy consumption, acoustics and so on;
- issues relating to people movement – such as security, means of escape, crowd handling and so on;
- the logistics of construction – such as where materials can be stocked on site to avoid multiple handling, damage and so on;
- safe access for construction and building maintenance;
- a solid basis for calculating whole life cost; and
- demonstrating effective operation of the completed facility.

Looking further ahead, where data is reliable, then it can also increasingly be asked to check itself, and applications are already available that will carry out design checks for compliance with building regulations, technical standards and so on. We should therefore look to the day when this is not just possible, but also the norm by which compliance (an entirely necessary but essentially non-productive and expensive process) is assured.

Surveys show that the potential business benefits of BIM are broadly accepted, with only between 10% and 20% of those surveyed responding that they do not believe that it will catch on or that it will add value. That itself would be a worrying statistic if it were not for the fact that it is always possible to find between 10% and 20% of people who will disagree with anything, so we can take the response as being overwhelmingly favourable – although some, particularly in the SME range of the market, are concerned about cost.

As for the suggestion that BIM should be re-named, maybe we should get to that when it is the biggest problem that we face, but in the meantime, even those who know very little about BIM know that it is something that they should know more about, so it would be best not confuse things by giving it a new name that may, or may not, refer to the same thing – not least as some protection against initiative fatigue.

The role of government and its BIM strategy

Much of the current energy in the BIM transformation programme was the result of an announcement made in November 2010 to the effect that Government would be requiring all of its projects to be delivered through 'Level 2' BIM by 2016. On its own, of course, such a statement, however bold, could not activate the change programme that it implied, but it was set as an objective precisely because other forces seemed to be coming together, as ripples on the ocean come together to create a wave.

Those factors include an improvement in the technology – the consequent use of that technology in some leading edge businesses, a few of whom are putting

Part III

BIM-based thinking at the core of their business operation – probably above all the impact of a recession that demands greater value in the delivery of construction projects, and as a consequence of all of the previously mentioned, some regular clients of the industry demanding both a better offer and an improvement on traditional, unreliable ways of doing business.

Government is one of those clients. The strategic importance of the industry in creating the means of production, commerce, shelter and social services, and the infrastructure that supports all of those, is unchanged, but everything has to be done for less money than in the past if it is to be affordable.

The driving idea around the adoption of BIM by Government is therefore that, whilst there are some immediate gains from the use of the technology, as summarised previously, the real gain will come from structural change in the industry and its practices. It is therefore an integral part of a wider Government Construction Strategy that aims to improve value for money for the industry's clients, including the Government itself [2], and the main change that is desired is integration, or Integrated Project Delivery – that is, for designers and their co-consultants, construction managers, specialist trade contractors, product manufacturers and facilities/asset managers to work together to find better ways of doing things. This should unlock innovation in a way that competition on price alone has failed to do. BIM therefore provides the means by which all of those team members can collaborate on the basis of shared common data, and it consequently does not just facilitate integration, but also it positively demands it if its full potential is to be realised.

Within that integrated team, however, every member will have slightly different requirements of BIM, and different risks and opportunities inherent in its use. Each company in every part of the industry will therefore have its own 'journey' to embed BIM in its own business, whatever that might be, and many will want to compete on the basis of an offer that is better than its competitors.

There is, however, no point in anyone having an offer that is so differentiated that there is an immediate loss of the fundamental benefit of BIM of all parties working on a common platform. Government has therefore taken a leadership role, not just in preparing itself and its suppliers for a world in which all projects are handled in BIM, but also in shaping how BIM itself might operate in practice and how the industry might work together to make it happen.

Looking at the last of these first, the Government has established a BIM Task Group, which brings together expertise from Government, the wider public sector, industry, institutions and academia. The Task Group has in turn motivated the creation of a network of 'BIM4 ... ' groups designed to develop the application of BIM in particular sectors (such as water, rail/infrastructure in general, retail, other private sector commercial development, etc.) and to disseminate information regionally (the Regional Network) and to specific industry constituencies (such as the supply chain generally, SMEs, BIM for FM and the BIM 2050 group for younger members of the industry). Further details of all of these are available on the Task Group website,[3] which is also a useful way of keeping up to date in this fast-changing landscape.

As far as the Government shaping its own programme is concerned, the first step was the publication of a high-level strategy for BIM.[4] This sets out the expectation for the adoption of BIM on centrally procured Government projects. Much of it is technical, but it also has a number of underlying principles, of which the primary ones are as follows:

- that it is agnostic: there should be no mandate that assumes the use of any proprietary product or system;
- that the many complex decisions in how BIM will be used in and between different businesses should stay in the supply chain, rather than become a matter for resolution by the industry's clients; and
- that these principles will allow innovation, but that there nonetheless needs to be sufficient convergence of practice to encourage people to invest without the fear of being 'beta-maxed' – that is, having backed the wrong technology – that, as complexity itself is an enemy of adoption, nothing should be complicated beyond the point at which it delivers widespread value.

There is another hope that cannot be assured if the market is to be left free to innovate, but that nonetheless remains an aspiration. That is that, whilst there might be a limited number of software vendors that provide the common platform that underpins a project model, beyond that it will be an 'app world', so that the host of specialists who are involved in a project can demand or develop applications designed for their own part of the task.

The final key principle of the Government strategy for BIM is that it should not be based on an assumption that technology problems that have not been solved to date will definitely have been solved by some fixed date in the future. An analogy would be that, when John F. Kennedy said that America would put a man on the moon by the end of the 1960s, he already knew that it was possible. It would not, therefore, be wise to develop a strategy that requires a new rocket (or its BIM equivalent) to be invented so that progress can be maintained. This does admittedly put a constraint on ambition, but the objective is not to blaze a trail at the leading edge, but rather to encourage mass adoption of this new approach, and this idea would be stillborn if it involved encouraging widespread investment in technologies that subsequently fail.

By getting involved to this extent, in what is fundamentally a matter for the industry, Government is not just seeking a better offer for the taxpayer in the purchase of public sector works, but also hopes that, as the industry's major client, it can motivate both reform and convergence in practice so that businesses can invest with confidence in the expectation of a steady, managed programme of change.

The real work will, however, have to be done by the industry, and every business will need to decide how it is going to enter the market, in terms of choosing the available technology most suited to its customers and fellows in the supply chain; investigating the changes that the adoption of BIM will demand of its existing policies; setting new standards, allocating resources and training staff; and considering how a distinctive approach towards the adoption of BIM can bring competitive advantage.

To consider the practicalities of widespread adoption, however, it is first necessary to look at the expectations of the current strategy in the context of a maturing technology and what this means in practice for businesses working in the mainstream of the construction industry.

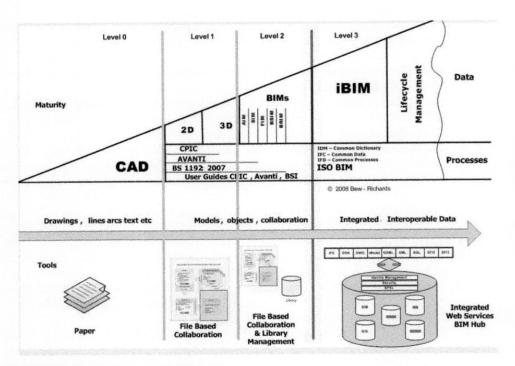

Figure 21.1   BIM maturity diagram.

The levels of BIM adoption

What the Government strategy says is that Government will require all of its projects to be managed in BIM at Level 2 by 2016. A shorthand definition of Level 2 BIM might be that, in accordance with the 'man on the moon' analogy, it comprises those things that we already know how to do – but that will nonetheless secure significant benefits.

It is, however, made clearer by a maturity diagram drawn by Mark Bew, the lead author of the strategy, and Mervyn Richards (see Figure 21.1).

In broad terms, the progression runs as follows:

- **Level 0** represents the early days of Computer-Aided Design, conducted in two dimensions, and with information exchanged either on paper or by handing over disks (this was before the days of email), and generally with no access by one team member to the work of another. It therefore represented the first stage of moving the design process off drawing boards and into computers. There were very limited standards, with the main means of control/integration being through the shared use of particular proprietary systems.
- **Level 1** reflects the increasing move from two dimensions into three, and the beginnings of file-based collaboration, but still with very limited access by one team member to the work of another. It also marks the point at which the need for industry standards was recognised, and standards started to be developed.

- **Level 2** moves to the use of three-dimensional modelling, progressing from line-based drafting (which essentially replicates the pre-digital process) into the creation of three-dimensional information-rich objects, with data exchanged by converting information to a commonly accessible format.
- **Level 3** envisages that there will be much more sharing of the three-dimensional model itself, with data being more readily exchanged without conversion, and also with the model being 'live' so that one team member can actually change the work of another.

The BIM journey

The principal differentiation between stages is therefore how completely individual team members can work on a single model, with open standards and a high degree of collaboration. But to get to this ideal, some key enablers need to be in place first, including the following:

- an agreed plan of work, which goes beyond current existing standards and sets out in more detail the deliverables required of each member of the team at defined stages which, in turn, means that both the deliverables and the stages need to be agreed;
- consistent terminology (or language), so that an agreed term means the same thing whenever it is used;
- codes by which that language might be used to build up libraries of descriptions for individual objects;
- either totally compatible software or a means of bringing different programmes together;
- agreed means by which all of the previously mentioned can be carried into the operation and maintenance phase of a project;
- co-ordinated terms of appointment by which all of those engaged in the process can be commissioned; and
- changes to other business practices and services (most particularly insurance) to support new responsibilities and inter-relationships.

In addition to these technical building blocks, there also needs to be a massive programme of cultural change by which institutions and individual businesses create a presumption of working together towards achieving aligned goals.

However, there is still a requirement for a strong technical foundation, with these key enablers implemented to an agreed standard, and a broad consideration of each follows.

Plan of work, deliverables and work stages

As already noted, BIM is fundamentally about the management of information. This means each party being clear about the information that it needs at every stage of the process and about the information it will provide. Achieving

Part III

agreement about that is clearly an iterative process, but one necessary outcome of that process has to be an agreed definition of what those stages are.

The timing of information is particularly relevant, and once again, the adoption of BIM, and processes suited to it, provides an opportunity of looking afresh at the information requirements of all of those involved in the development process.

The early experience of BIM is that, in order to obtain optimal benefit of the use of the technology, more players are involved in the process earlier in the development period, and because it is both necessary and possible to resolve some design problems earlier, design work becomes more 'front-end loaded' than under more traditional pre-construction processes. This calls for a new plan of work, which not only reflects this bringing forward of design operations, but also goes into greater detail about the information (both spatial and technical) that each member of the design process needs to produce at each work stage.

This alone – the idea of all consultants working on a single plan by which the work of all of them is defined on the same principles – is a considerable prize by contrast with the status quo, where the terms of reference of separately appointed consultants are rarely comprehensive or well aligned.

The professional institutions are therefore working on a revised plan of work that will reflect the necessary differences in planning a project to be administered in BIM, and the RIBA has produced a 'BIM Overlay' to its Plan of Work that represents progress in that direction.[5] Greater detail is required, however, and this will be incorporated in an Information Delivery Manual ('IDM'), which will effectively become the Plan of Work.

In the meantime, the BIM Task Group has developed a standard describing how BIM works at Level 2 in the project delivery phase.[6]

Current thoughts about the stages (or 'data drops') at which these defined sets of information might be delivered, generally relating to groups of stages in the RIBA plan of work, are as follows:

- Drop 1: Requirements and Constraints (with simple 'boxes' used to represent individual spaces in the model, and with Room Data Sheets representing functional, environmental and finishes requirements).
- Drop 2: Outline Solution (with the first representative three-dimensional model, incorporating MEP and FF&E items generically, and with Room Data Sheets now representing the design solution in terms of function, environmental conditions, finishes and FF&E).
- Drop 3: Construction Information (with the model developed so that MEP and FF&E items are now shown as specific products, and with Room Data Sheets showing the requirements for environmental conditions, finishes and FF&E on a basis adequate for construction).
- Drop 4: Operations and Maintenance Information (as data drop 3, but now showing 'as built' conditions, and adding requirements for operations and maintenance).
- Drop 5 (and thereafter): Post-occupancy validation information and ongoing operation and maintenance.

There then needs to be an equally clear understanding of what is to be delivered at the beginning of each work stage (principally in terms of briefing information from the client), giving all team members a basis for progressing the design (and later the construction and building operation), and also the information that will be required at the end of the work stage (deliverables) to give the client the information needed to make the decisions that are required in order to proceed to the next stage. This too will be set out in the IDM.

Loading the model: language and libraries

A particular feature of BIM (as opposed to just three-dimensional CAD) is that 'objects' in the model itself can be loaded with information – whether they might be individual products, or assemblies, or whole elements, or spaces within the building.

In some ways, product data offers a metaphor for the change that the full implementation of BIM will drive through the industry. For some time now, the use of a printed brochure has slowly been overtaken by product information being conveyed electronically, ideally online. BIM, however, will not just render the printed brochure obsolete, it will also demand far more than 'dumb' electronic information. Rather, it will require information about every product, assembly or element to be presented as a three-dimensional object, loaded with the information necessary to incorporate it into the design, execute the work on site and then operate the asset.

The two qualities most talked of in this context are cost and carbon, but such information might also include, for example, the following:

- for the designers: functional performance, the necessary means of support, any connections to services, interfaces with other elements of the design and so on;
- for the contractor: weight, any special handling arrangements, the need for protection prior to handover and so on;
- and for the building owner: energy consumption, maintenance or replacement requirements and so on;

and much more besides.

Again, exactly what information is required will be the result of a combination of the industry developing protocols as a baseline standard and then individual businesses providing more as part of a competitive offer.

At high level, these qualities can be characterised as a further dimension of BIM, moving beyond three dimensions to 4 (adding time), 5 (cost), 6 (life cycle cost) and so on. It is almost inevitable that companies will seek to compete around the addition of further dimensions, but the critical thing will be to make BIM work properly in three dimensions first, and then agree codes and protocols by which additional dimensions might be added in a way that adds value.

It follows that each individual profession will also need to work out exactly what information might be useful and how it might be used. For example, for the quantity surveyor, including the cost of each element of the model might make it possible (at least in theory) to build up the total project cost as elements are loaded into the model. The reality will be more complex, however. Even at product level, cost is not a fixed concept: the cost of a component will vary as it passes through the supply chain, and even the ex works price will vary depending on the quantity purchased, the quality of the relationship with the supplier and so on. Similar complications will apply to other properties, so once again, this points to the need for the industry to agree standards and protocols by which such information will be developed and communicated.

However, these complexities have to be managed whether working in a BIM model or with a quill pen, so they are not obstacles to progress, and there will only be two constraints over the amount of information that can be loaded into the model, and the number of applications that can be mobilised to use it: our imagination and utility – the discipline required to ensure that models are not overloaded with information and complexity that does not produce value in use. As with so many other things in life, just because we can do something doesn't mean that we should.

Such objects, loaded with their data, can also be rolled forward from project to project, addressing another historic problem of the industry – its tendency to re-design the wheel from project to project.

To assist the development of objects for both their original use in a project and then their repeat use also requires, however, that they are produced in accordance with generally adopted standards. Another current piece of work being under-taken by the BIM task groups is therefore the agreement of product (or object) libraries, which codify the way that the data relating to a particular object is built up. This work has already advanced sufficiently far to be in use on the Pathfinder projects commissioned by Government.

Bringing different software programmes together – the search for interoperability

A key word in the context of BIM, and in considering the progression through the levels of maturity shown in Figure 21.1, is 'interoperability' – which may prove as difficult to achieve as it is to pronounce. This is the holy grail of finding a means by which different systems or models can communicate with each other. Progress is being made on this front – or more accurately on several fronts, as there is no consensus as to the best way forward in the search for a neutral, open system of exchange that does not suffer loss of fidelity in the base data. It is a search that has been in progress for almost 20 years already, and it may be that, similarly to the holy grail, its goal is not totally achievable, and that rather than find a system by which any one software vendor's product can 'talk' to all others, there may be a more selective system by which bits of data that do actually need to connect can find each other.

Either way, the need for interoperability in some form becomes more and more important as BIM adoption progresses through the levels, and teams seek to work more and more collaboratively, making greater use of 'live' shared data.

The only practical alternative to a neutral, open means of exchange is for a single vendor system to be mandated. This may be possible for individual sup-ply chains, but it is not something that can or should be mandated by Govern-ment because of the stifling effect on what might otherwise be an innovative marketplace; mandating single systems creates a problem when supply chains are constantly re-forming, so that individual companies may be on the receiving end of a mandate to use quite different systems from different clients or teams.

A generally accepted means of exchange would therefore be a better alternative. The two most favoured current systems are as follows:

- The Industry Foundation Classes (IFC) date standard under continuous development by building SMART – the successor organisation to the Industry Alliance for Interoperability established in 1994. IFC is a neutral, open specification (i.e. not controlled by a single vendor or group of vendors) intended to describe building and construction asset data. It is currently in the process of becoming the official International Standard ISO 1673916.
- The Application Programming Interface (API) exchange systems developed by vendors.

There is, however, some way to go before either or both are accepted by IT specialists in the industry as the answer to the interoperability problem.

In the meantime, the tool that is most used to exchange data between people who may be operating on different systems, and which is embedded in the Government strategy for BIM, is COBie – the Construction Operations Building Information Exchange. This is a system developed by the Army Corps of Engineers in the United States, and available (as COBie2, in a version localised for use in the United Kingdom) free to users under licence. It was originally intended for organising the handover of record documents and maintenance manuals at project completion, but it serves as a means of exchanging data in a structured way at any stage of the project. Critically, it is also simple, with the actual exchange made using nothing more complex than a spreadsheet – although this obviously means that the graphics are lost. However, this information can then be converted back to systems operated by designers or by facility managers.

It is the use of COBie that has made it possible to make real progress in the use of collaborative working within a BIM environment in advance of all of the problems of interoperability being solved, and it is likely to be the standard means of data exchange between systems until at least 2016, when the Government strategy moves on towards Level 3.

It is therefore important for practitioners to understand how COBie works, so that data is entered (by designers, product manufacturers and contractors) as and when it is required. It can also be anticipated that there will be improvements in the software available to make it easier to translate data into and out of COBie format.

Operation and maintenance

Over the last few years, the Building Services Research and Information Association (BSRIA), working with the Usable Buildings Trust, has developed a methodology for reflecting in the design process the real requirements of the subsequent operation and maintenance of a facility.

This methodology, called 'Soft Landings', addresses the too-frequent gap between the design expectation and actual performance of a built asset. It does so by requiring clients (and critically those who will manage the completed facility),

Part III

designers and constructors to spend more time on early constructive dialogue; setting achievable targets at commencement; preparing for and managing handover, occupation and use; and continuing the process into aftercare to get the best from the asset, including post-occupancy evaluation and feedback into evidence-based design.

Soft Landings is therefore being incorporated into the Integrated Design Manual as part of the BIM adoption process, addressing a long-standing defect in the UK design and construction process, which is the absence of an effective feedback loop.

Terms of appointment and changes to other business practices

Despite the general consensus about the benefits of BIM, and the positive return on investment, it is not surprising that there remain obstacles to the introduction of potentially so radical a change in current practice. Many of these relate to concerns about new terms of trade, and these concerns include the following:

- the protection of intellectual property;
- accountability, and how liability for negligence can be established when many people have access to, and the ability to change, the model;
- possibly new or additional liabilities created by accepting responsibility for managing the model;
- as a consequence of the previously mentioned, the need for the insurance industry to get comfortable with (probably new) ways of insuring the professional liability of all participants;
- holding information secure, whilst keeping it accessible for those who need it for the purposes of design and construction;
- system capacity necessary to store and transmit very large documents; and
- the development of standard terms of appointment appropriate for an integrated team working in a BIM environment.

As noted, for Level 2 BIM, the approach to these problems has been to seek the minimum necessary change from current ways of working. For example, on the potentially vexed question of 'who owns the model?', the short-term answer is that there is no need for great change to the current practice – which is that the producers of designs retain the copyright, but the client has an absolute right to use that information for all legitimate purposes connected to ownership of the resulting asset. Issues will, however, arise (if only between team members, rather than between the client and the team) as they work more collaboratively on a shared model, and those problems that are being by-passed at Level 2 will need to be fully resolved before progressing to the widespread adoption of Level 3 BIM.

For example, whilst not every member of the team needs to have access to 'live' design information produced by all of the others, there does need to be assurance that changes made by one team member that affect others are brought to the model in a way that ensures the impact on other elements of the design is picked up.

The facility for information to be shared and coordinated in this way is a critical advantage of BIM, as it matures beyond the unstructured use of CAD.

Level 3 and the future

Looking further ahead, it remains the objective of the Government that by 2016 all Government projects will be handled in Level 2 BIM, but also that, by that time, it should be possible to commence early adopter projects in Level 3.

This will call for a fully functioning technology for more direct exchange of information between systems, and for some reconsideration of the legal protocols by which we operate. Many other practical details will change as well, as BIM development moves into Level 3, but it is also critical to look ahead to the way technology will be used in 2016 and the foreseeable future beyond that, and it is already anticipated that there will be significant changes in current practice. These will include a migration away from the desktop towards more handheld devices; away from the dominance of current operating systems; away from terminals loaded with software towards working in the cloud; away from information being transferred from private terminal to private terminal towards transactions being conducted live online; and with all of the previously mentioned taking place against a background of a geometric increase in the amount of data that is being handled.

Each of these changes will open up new challenges, but also new possibilities – not just in the use of technology (e.g. with greater use being made of handheld appliances out in the field), but also in industry processes. As just one example of this, if projects are available online, then instead of procurement being a one-way process by which building owners or their agents invite prices for designed objects or pieces of work, suppliers can come to the client and offer products that fit their specification, or offer different and better ways of achieving the same result.

In a similar way, and given the right programme, pre-designed objects, with all of their information, might be suggested to designers either from a published library or from their own previous work, together with guidance on how the object might be incorporated into a new design.

The sharing of data that is exchanged in digital form will open up almost infinite possibilities and will effect far-reaching changes in the structure and practice of the industry. Again, the only limitation is our collective imagination tested against the need to add value.

Epilogue

So to finish as we began: BIM is about much more than software. It is about unlocking the potential for the more imaginative actors in the industry to find new ways of doing business to produce a better price offer and better long-term value

Part III

for their clients. Similarly to any game-changing initiative, therefore, it represents both an opportunity and a threat: opportunity for those who see and embrace its potential, and threat for those who may see it either as a passing fad or as a good idea for other people.

Notes

1. www.cpic.org.uk.
2. Government Construction Strategy, May 2011 (www.cabinetoffice.gov.uk/sites/default/files/resources/Government-Construction-Strategy.pdf).
3. www.bimtaskgroup.
4. BIM Management for Value, Cost and Carbon Improvement, March 2011 (http://www.bimtaskgroup.org/wp-content/uploads/2012/03/BIS-BIM-strategy-Report.pdf).
5. BIM Overlay – the RIBA Outline Plan of Work, May 2012. (www.ribabookshops.com/uploads/b1e09aa7-c021-e684-a548-b3091db16d03.pdf)
6. PAS1192 – 2:2012: Building Information Management – information requirements for the capital delivery of construction projects. (www.bimtaskgroup.org/pas-1192-22012).

Chapter 22
Bills of Quantities

Alan Muse

Tender and contract document

Bills of quantities are documents that describe the quality and give the quantities of the constituent parts of proposed building works. They have two primary functions. Initially, they are used as tender procurement documents to provide a uniform basis for competitive lump sum tenders. Subsequently, they become contract documents serving as schedules of rates for the pricing of variations. It is therefore important that they contain at least the basic information required of them by the prevailing conditions of contract and that they are presented in a recognisable format that will facilitate their use.

The SBC With Quantities requires bills of quantities to be prepared in accordance with the RICS New Rules of Measurement (Detailed Measurement for Building Works: NRM 2).

The wider role

Even with the more or less universally adopted use of computers, the cost of producing bills of quantities is high, and consequently, their use as tender procurement and contract documents is often questioned and new techniques and procedures aimed at supplanting them are often tried. However, bills of quantities have survived thus far, no doubt because in addition to their two primary functions, they contain vast amounts of information that can be of use in many ways. Also, they are frequently used in the preparation of interim valuations (see Chapter 30).

In preparing bills of quantities, the quantity surveyor sifts through and processes much detailed information that, although not required by NRM 2 to be included in the bills of quantities, may be of use to tendering contractors and

The Aqua Group Guide to Procurement, Tendering and Contract Administration, Second Edition.
Edited by Mark Hackett and Gary Statham.
© 2016 John Wiley & Sons, Ltd. Published 2016 by John Wiley & Sons, Ltd.

the project team in connection with project planning, administration and the compilation of historical cost data in the form of elemental cost analyses. It is therefore worthwhile, in the early stages of the plan of work, to consider what additional information can be readily gathered in the measurement process and/or presented in the bills of quantities to assist tenderers and the project team to operate more effectively and efficiently.

Basic information

Bills of quantities are prepared by describing and quantifying, in accordance with the rules of NRM 2, the work shown on the drawings and given in the specification. Amongst other things, NRM 2 requires that 'Bills of Quantities shall fully describe and accurately represent the quantity and quality of the works to be carried out'.

It is normal practice to arrange bills of quantities in three generic sections that between them fully describe and accurately represent the quantity and quality of works to be carried out (NRM 2, 3.1.3). These are preliminaries, preambles and measured works.

Preliminaries

NRM 2 contains rules for describing and quantifying preliminaries. The aim is to provide a comprehensive schedule of items relating to the project, generally, the contract, the 'employer's requirements', the contractor's general cost items and work by others that can be individually priced by contractors for the purposes of calculating their lump sum price and for fixing rates that can be subsequently used in the valuation of variations.

Work Section 1 comprises the rules for describing and quantifying preliminaries. It is divided in two sub-sections as follows:

- Preliminaries (main contract); and
- Preliminaries (work package contract).

Both sub-sections are sub-divided into parts as follows:

- Part A: Information and requirement (i.e. dealing with the descriptive part of the preliminaries); and
- Part B: Pricing schedule (i.e. providing the basis of a pricing document for preliminaries).

Provisional sums are defined in both the SBC and NRM 2. They are used to make cost provision within the bills of quantities for works to be carried out by local authorities or statutory undertakers, dayworks and other defined and undefined provisional work that cannot be described and given as items in accordance with the rules of NRM 2. Also, the SBC requires the architect or contract administrator to issue instructions in regard to the expenditure of provisional sums.

Rules 2.9.1.1 to 2.9.1.6 of Part 2 of NRM 2 clearly set down what is meant by defined and undefined provisional work in clause 1.6.3 (Definitions), and these rules are referred to in the definitions section of the SBC with quantities. Provided that all of the required information is given in the description of defined provisional work attached to a provisional sum, then the contractor is deemed to have made due allowance in his tender for that work in programming, planning and pricing preliminaries. However, where all of the required information is not given, the work is deemed to be undefined provisional work, and the contractor will be deemed not to have made any such allowance. It is therefore of utmost importance that provisional sums for undefined provisional work are set at a level that covers the costs deemed not to have been included by the contractor in his lump sum price. It is also worthy of note that, when the naming of a supplier or sub-contractor or work by local authorities or statutory undertakers occurs as a result of expenditure of a provisional sum, that sum will have to be sufficient to also cover the cost of the main contractor's profit and attendances, and the fixing only of any supplied items.

Preambles

The preambles traditionally comprise clauses specifying the quality or standard of materials and workmanship required. However, the preambles now more commonly comprise a simple reference to a separate, comprehensive specification document (see Chapter 20), which is thereby incorporated into the bills of quantities and given contract document status. Either way these specification requirements are intended to reduce the need for long descriptions and repetition in the measured works section of the bills of quantities. It is not normal for items in the preambles to be directly priced. The cost of complying with them is usually included within the rates set against the individual items included in the measured works section of the bills of quantities.

Measured works

The measured works section comprises a schedule of quantified descriptions of the constituent parts of a building project (including services installations and external works) compiled in accordance with NRM 2. Provided that the descriptions comply with the appropriate rules, the quantities may be approximated if they cannot be accurately determined at bill production stage. The items are normally arranged in the NRM 2 order of work sections, which closely follow the Uniclass referencing system. A cross-referenced index is provided to assist the surveyor in annotating the bill descriptions. The aim is to provide a comprehensive schedule of items, described and quantified according to industry recognised conventions, which can be individually priced by contractors for the purposes of calculating their lump sum price and for fixing rates that can be used subsequently in the valuation of variations.

Part III

Provided that the Uniclass classification referencing system is used to annotate the drawings, specifications and bills of quantities, items in the bills can be directly related through that referencing system to relevant items in the specification and details on the drawings.

Formats

For tender purposes, it is normal to arrange bills of quantities in the Uniclass work sections, as Example 22.1. However, it may be desirable, for various reasons, to add further information and/or to rearrange the format. Provided that the quantity surveyor appropriately annotates all items when measuring the work, the resultant bills of quantities can be readily reformatted to provide, for example, the following:

- **Locational bills of quantities**, as Example 22.2, in which the Uniclass work sections are still adopted but the quantity for each item is broken down and allocated to a particular position within the project (e.g. a particular building, a part of a building or a house type). This format was developed to assist in the more accurate pricing of items.
- **Annotated bills of quantities**, as Example 22.3, in which the Uniclass work sections are still adopted but each item is annotated as to what it is and where it is located within the project. The annotations can be provided in a separate document, in a separate section of the bill or, most usefully, facing each item on the back of the preceding page in the bill. This format was developed as an extension of the locational bill.
- **Elemental bills of quantities**, as Example 22.4, in which the sections are based on the functional elements of the building, normally those used by the Building Cost Information Service (BCIS) (see Chapter 18), with the work in each element being arranged in Uniclass work sections. This format was developed to facilitate elemental cost analyses and as an aid to cost planning.

Early contractor involvement is a trend in the industry. There are various ways that this can be achieved, but quite often, the project is split into discrete 'packages' that are let by the contractor as separate 'works contracts', 'trade contracts' or 'sub-contracts'. The precise number of packages and their demarcation are determined by the design team and the contractor on a project-by-project basis. The contractor's expertise and advice are relied upon to attain the most efficient way of splitting the project into packages. Care is needed in coordinating the interfacing of packages to ensure that items of work are neither overlooked nor duplicated. Complete bills of quantities may be prepared for each package using any or all of the formats described previously.

The examples of different formats for bills of quantities as given at the end of this chapter are not exhaustive. It is up to the design team in general and the quantity surveyor in particular to decide how best to format the bills of quantities for each project so as to maximise the benefit to all concerned. It is worth remembering that efficient management of documentation is a valuable aid to efficient working on site and that can be of cost benefit to the employer.

Part III

Note: extracts from two separate sections covering three trades are set out as follows to allow comparison with other formats. The sections and items within them are in the same order as NRM 2.

F MASONRY   £
F10 Brick/block walling

Concrete blocks, BS6073, 440 mm × 215 mm, solid, keyed one side, in cement-lime mortar (1:6), stretcher bond.

Walls; blockwork

A	100 mm thick	427 m^2
B	100 mm thick; curved on plan to 150 mm radius	12 m^2
C	200 mm thick	219 m^2

M SURFACE FINISHES
Floor, wall, ceiling and roof finishes
M20 Plastered/Rendered/Roughcast coatings

Plaster; BS1191 part 2, undercoat 11 mm thick; finishing coat 2 mm thick; on brickwork or blockwork

To walls, 13 mm thick; steel trowelled

H	>600 mm wide	860 m^2

Plaster; BS1191 part 2, undercoat 11 mm thick; finishing coat 2 mm thick; on concrete

To ceilings, 13 mm thick, steel trowelled

J	>600 mm wide	435 m^2

To isolated columns, 13 mm thick, steel trowelled

K	≤600 mm wide	22 m

M60 PAINTING/CLEAR FINISHING
Decoration

One coat sealer; two coats emulsion paint, matt finish; plaster base; roller applied

General surfaces

T	>300mm girth	1295 m^2
U	≤300 mm girth	22 m

Part III

Example 22.1   Uniclass work section bill of quantities.

Note: extracts from two different sections covering these trades are set out as follows to allow comparison with other formats. The sections and items within them are in the same order as NRM2. The quantity of each item is broken down and allocated to a particular position within the project here represented by the letters A, B and C.

F MASONRY   £
F10 Brick/block walling

Concrete blocks, BS6073, 440 mm × 215 mm, solid, keyed one side, in cement-lime mortar (1:6), stretcher bond.

Walls; blockwork

A	100 mm thick A 340 + B 67 + C 20	427 m²
B	100 mm thick; curved on plan to 1500 mm radius A 0 + B 57 + C 63	12 m²
C	200 mm thick A 99 + B 57 + C 63	219 m²

M SURFACE FINISHES
Floor, wall, ceiling and roof finishes
M20 Plastered/Rendered/Roughcast coatings

Plaster; BS1191 part 2, undercoat 11 mm thick; finishing coat 2 mm thick; on brickwork or blockwork

To walls, 13 mm thick; steel trowelled

H	>600 mm wide A 688 + B 129 + C 43	860 m²

Plaster; BS1191 part 2, undercoat 11 mm thick; finishing coat 2 mm thick; on concrete

To ceilings, 13 mm thick; steel trowelled

J	>600 mm wide A 350 + B 65 + C 20	435 m²

To isolated columns, 13 mm thick; steel trowelled

K	≤600 mm wide A 15 + B 0 + C 7	22 m

M60 PAINTING/CLEAR FINISHING
Decoration

One coat sealer; two coats emulsion paint, matt finish; plaster base; roller applied

General surfaces

T	>300 mm girth A 1038 + B 194 + C 63	1295 m²
U	≤300 mm girth A 15 + B 0 + C 7	22 m

Example 22.2 Locational bill of quantities.

Note: extracts from two separate sections covering three trades are set out on the facing page to allow comparison with other formats. The sections and items within them are in the same order an NRM2. Annotations are given as follows.

F MASONRY
F10 Brick/block walling

Annotations

A   Non-load-bearing partitions, first floor, (drawing 456/78)

B   Stores, ground and first floor and staircase enclosure walls

C   Non-load-bearing partitions, ground floor (drawing (456/77)

M SURFACE FINISHES
Floor, wall, ceiling and roof finishes
M20 Plastered/Rendered/Roughcast coatings

Annotations

H   Load-bearing partitions, ground floor

J   Soffit of first floor

K   Columns, entrance hall

M60 PAINTING/CLEAR FINISHING
Decoration

Annotations

T   Load-bearing partitions, ground floor and soffit of first floor (refer to colour schedule).

U   Columns, entrance hall (refer to colour schedule).

Part III

Example 22.3   Annotated bills of quantities.

Note: extracts from two separate sections covering three trades are set out on the facing page to allow comparison with other formats. The sections and items within them are in the same order as NRM2. Annotations are set out on the facing page.

F MASONRY   £
F10 Brick/block walling

Concrete blocks, BS6073, 440 mm × 215 mm, solid, keyed
one side, in cement-lime mortar (1:6), stretcher bond

Walls; blockwork

A	100 mm thick	427 m²
B	100 mm thick; curved on plan to 1500 mm radius	12 m²
C	200 mm thick	219 m²

M SURFACE FINISHES
Floor, wall, ceiling and roof finishes
M20 Plastered/Rendered/Roughcast coatings

Plaster; BS1191 part 2, undercoat 11 mm thick; finishing
coat 2 mm thick; on brickwork or blockwork

To walls, 13 mm; steel trowelled

>600 mm wide   860 m²

Plaster; BS1191 part 2, undercoat 11 mm thick; finishing
coat 2 mm thick; on concrete

To ceilings, 13 mm thick; steel trowelled

>600 mm wide   435 m²

To isolated columns, 13 mm thick; steel trowelled

≤600 mm wide   22 m

M60 PAINTING/CLEAR FINISHING
Decoration

One coat sealer; two coats emulsion paint, matt finish;
plaster base; roller applied

General surfaces

T	>300 mm girth	1295 m²
U	≤300 mm girth	22 m

Example 22.3 *(Cont.)*

Note: extracts from two separate sections covering three trades are set out on the facing page to allow comparison with other formats. The elements are set out in the same order as is used by BCIS; sections and items within each element are in the same order as NRM 2.

2G INTERNAL WALLS AND PARTITIONS
F10 Brick/block walling

£

Concrete blocks, BS6073, 440 mm × 215 mm, solid, keyed one side, in cement-lime mortar (1:6), stretcher bond.

Walls; blockwork

A	100 mm thick	427 m²
B	100 mm thick; curved on plan to 1500 mm radius	12 m²
C	200 mm thick	219 m²

3A WALL FINISHES
Floor, wall, ceiling and roof finishes
M20 Plastered/Rendered/Roughcast coatings

Plaster; BS1191 part 2, undercoat 11 mm thick; finishing coat 2 mm thick; on brickwork or blockwork

To walls, 13 mm thick; steel trowelled

H	>600 mm wide	860 m²

To isolated columns, 13 mm thick; steel trowelled

J	≤600 mm wide	22 m

M60 PAINTING/CLEAR FINISHING
Decoration

One coat sealer; two coats emulsion paint, matt finish; plaster base; roller applied

General surfaces

K	>300 mm girth	860 m²
L	≤300 mm girth	22 m

3C CEILING FINISHES
Floor, wall, ceiling and roof finishes
M20 Plastered/Rendered/Roughcast coatings

Plaster; BS1191 part 2, undercoat 11 mm thick; finishing coat 2 mm thick; on concrete

To ceilings, 13 mm thick; steel trowelled

T	>600 mm wide	435 m²

M60 PAINTING/CLEAR FINISHING
Decoration

One coat sealer; two coats emulsion paint, matt finish; plaster base; roller applied

General surfaces

U	>300 mm girth	435 m²

Part III

Example 22.4   Elemental bills of quantities.

Chapter 23
Sub-contractors

Andrew Shaw

Introduction

The SBC is drafted on the premise that the contractor will be wholly responsible for carrying out and completing the works. However, as sub-contracting is customary in the construction industry, the SBC makes provision for such arrangements by recognising three basic categories of sub-contractors to whom a part of the works may or must be sub-let: there are those sub-contractors that the contractor uses of his own volition (commonly referred to as 'domestic sub-contractors'); those sub-contractors that the contractor is directed to choose from a select list provided by the employer (commonly referred to as 'listed sub-contractors' or 'named sub-contractors'); and those sub-contractors that the employer requires the contractor to use for any particular package of work (referred to as 'named specialists').

- **Domestic sub-contractors** – Provided that the contractor obtains the written consent of the architect/contract administrator, which consent must not be unreasonably delayed or withheld, he may sub-contract work to any person of his choice, and such a person is referred to as a domestic sub-contractor.
- **Listed sub-contractors** – The employer may require certain items of work to be executed by one of not less than three persons named in a list within the contract documents. In such a case, the contractor must sub-contract that work to one of the listed persons, and such person is also referred to as a sub-contractor.
- **Named specialists** – Under the optional provision of the 2011 'Named Specialist Update', the employer can require the contractor to employ a particular person for works identified in the contract (either a pre-named specialist or, for works under a provisional sum, a post-named specialist).

It is important to note that the contractor remains wholly responsible for carrying out and completing the work whether or not it is sub-contracted.

The Aqua Group Guide to Procurement, Tendering and Contract Administration, Second Edition.
Edited by Mark Hackett and Gary Statham.
© 2016 John Wiley & Sons, Ltd. Published 2016 by John Wiley & Sons, Ltd.

Specialist sub-contractors

As buildings become more complex and the systems within them more sophisticated, early decisions have to be made on various elements of them before the design team is able to proceed with the detailed design. For example, much thought must be given to the detail of a structural frame before the cladding of it can be designed; mechanical and electrical systems are frequently complex and make considerable demands on space for both plant and distribution that can affect storey heights and floor layouts. Thus, it is essential for decisions relating to a number of functional elements of the building, including the selection of the principal specialist sub-contractors, to be made very early in the design process. Sub-contractors selected in this way are invariably required to contribute to or be responsible for some of the detailed design. This frequently happens in the case of structural steelwork, external wall cladding, mechanical and electrical systems, lifts and escalators.

This early involvement of sub-contractors in the design process creates a special relationship between them and the employer and, in due course, leads to the situation in which the employer requires the contractor to enter into a contract with the selected sub-contractor. The contractual arrangements for this type of sub-contract require careful consideration.

Early JCT standard forms of contract made special provisions for sub-contractors selected (nominated) in this way. Unfortunately, those provisions had many shortcomings, in terms of both the contractual relationship of the parties and the responsibility for any design input, giving rise to disputes that often led to arbitration and litigation. Although the JCT refined its procedures for nominating sub-contractors through standard forms and tendering procedures, employers and their advisors used to see the allocation of risk to the employer as a significant and unjustified burden. The use of the nominating procedures in JCT standard forms diminished to such an extent that they were omitted from the SBC when the JCT revised JCT 98. The 'named specialist update' provision allows for a new form of nomination, which contains much simpler contractual relationship than was previously the case. There is no longer a separate contract between the specialist and the employer and which was linked to that of the contractor. The contractor meanwhile is afforded some protection from the employer where: (i) the named specialist become insolvent; and/or (ii) the contractor makes a reasonable objection to the post-named specialist. In both cases, the contractor is entitled to an extension of time and recovery of loss and/or expense.

Design by the sub-contractor

The SBC contemplates that part of the design will be undertaken by the contractor. Where there is a requirement for the contractor to design part of the works, the SBC contains provisions for a Contractor's Designed Portion (CDP). Where the contractor is required to design the whole of the works, the appropriate form of JCT contract to use is the Design and Build (DB) Contract.

Part III

Whether the main contract contemplates design by the main contractor, design input is frequently required from sub-contractors. Where such design input is properly planned, co-ordinated with the other production information and made available to the main contractor on time, as good practice dictates, there are seldom any practical problems. However, when this does not happen and disruption and/or delays result from a deficiency in the sub-contractor's design, the employer is likely to suffer loss of time and money. The employer's remedy is to recover any such losses from the main contractor, who would undoubtedly seek redress from the sub-contractor concerned.

The SBC and sub-contract agreements

The JCT produces three Standard Building Sub-Contracts for use with each of the three versions of the SBC (SBC/Q, SBC/AQ and SBC/XQ). The short form of sub-contract ('ShortSub') is for small scale works that are simple, straightforward, low risk and have no requirement for sub-contract design. Of the 'long' sub-contract forms, one version (the SBCSub) does not contain provisions for sub-contract design and the other version (the SBCSub/D) does, via a 'sub-contractor's designed portion'. The SBCSub/D design provisions are much the same as those under a CDP in the SBC.

Each of the two 'long' sub-contracts comprises two documents:

- SBCSub comprises:
  - SBCSub/A the agreement
  - SBCSub/C the conditions

- SBCSub/D comprises:
  - SBCSub/D/A the agreement
  - SBCSub/D/C the conditions

The agreement document to each of the two contracts records the variable parts of the contract such as the articles of agreement, recitals, articles, sub-contract particulars, attestation, schedule of information and supplemental particulars. As one might glean from the aforementioned list, and in contrast to the SBC and the short form of sub-contract, which are single documents, the SBC sub-contracts are split into two documents, one document being the formal agreement and the other being the contract conditions. The agreement document covers the articles of agreement, recitals, attestation and specific contract variables, such as the sub-contract particulars, schedule of information and supplemental particulars. As one might expect, the contract conditions document records how the contract should be administered.

When using SBCSub/D, it is integral to the agreement that the sub-contractor warrants to exercise reasonable skill and care in the:

- design of the sub-contract works;
- selection of materials and goods for the sub-contract works; and
- satisfaction of any of the contractor's requirements.

The limitation of the sub-contractor's design liability is, broadly speaking, the same as that limiting a main contractor's liability under a CDP of the SBC. Clause 2.13.1 of SBCSub/D/C 2011states that:

'the Sub-Contractor shall in respect of any inadequacy in such design have the like liability to the Contractor, whether under statute or otherwise, as would an architect or, as the case may be, other appropriate professional designer holding himself out as competent to take on work for such design who, acting independently under a separate contract with the Contractor, has supplied such design for or in connection with works to be carried out and completed by a building contractor who is not the supplier of the design.'

SBC provisions under the main contract

The SBC provides for a number of mandatory conditions that all sub-contracts must contain. These, broadly speaking, comprise requirements that:

- The sub-contractor's employment under the sub-contract must terminate immediately if the main contractor's employment is terminated under the main contract.
- No unfixed materials and goods delivered to, placed on or adjacent to the works and intended for use in the works shall be removed without first having obtained the main contractor's written consent to such removal, and the main contractor must himself receive consent from the architect/contract administrator before consenting to the removal of such materials and goods.
- Where the value of any unfixed materials and goods have been included in an interim certificate for payment to the main contractor and paid to him, those materials and goods shall become the property of the employer and the sub-contractor shall not dispute this.
- Where the main contractor pays the sub-contractor for any unfixed materials and goods before their value is included in an interim certificate, upon payment those materials and goods shall become property of the main contractor.
- The sub-contractor shall grant the rights of access to workshops or other premises where work is being prepared for the contract to the architect/contract administrator or any person authorised by him.
- If the main contractor fails to pay the amount properly due to the sub-contractor by the final date for payment, the contractor shall, in addition to the amount properly due, pay simple interest there on.
- Where applicable, the sub-contractor must deliver a warranty in favour of the employer within 14 days of receiving a written request from the contractor. This is intended to give the employer/funder/purchaser/tenant a right of recourse against the sub-contractor (for the sub-contractor's obligations) should the main contactor cease to exist. The JCT publishes four standard sub-contractor and trade contractor warranties for a number of its 2011 suite of contracts for the funder (SCWa/F), purchaser and tenant (SCWa/P&T and TCWa/P&T), and the employer (SCWa/E).

Part III

Chapter 24
Obtaining Tenders

Introduction

It is frequently desirable to appoint the contractor at an early stage in the design process, but to do so necessitates various modifications to traditional tendering procedures (see Chapters 4 and 17). Having said that, traditional tendering procedures are still widely adopted and are therefore considered in more detail in this chapter.

There have historically been a number of codes relating to the selection of contractors, of which the two most significant were as follows:

- The National Joint Consultative Committee for Building (NJCC) *Code of Procedure for Single Stage Selective Tendering* published in collaboration with the Scottish Joint Consultative Committee and The Joint Consultative Committee for Building, Northern Ireland (published 1996).
- The Construction Industry Board (CIB) *Code of Practice for the Selection of Main Contractors* (published 1997).

The NJCC and CIB no longer exist, and their codes are no longer published.

In July 2002, the JCT published *Practice Note 6 Main Contract Tendering (series 2)* to provide model forms of tender.

More recent guidelines are set out in the NBS *Guide to Tendering: for construction projects* (first published in February 2011) and which replaces the NJCC code. This guide is based on the Public Contracts Regulations 2006 ('PCR 2006'), which itself was enacted in order that UK law reflected recent consolidation of European Union procurement law. PCR 2006 has since been amended in part by the Public Procurement (Miscellaneous Amendments) Regulations 2011.

The Aqua Group Guide to Procurement, Tendering and Contract Administration, Second Edition.
Edited by Mark Hackett and Gary Statham.
© 2016 John Wiley & Sons, Ltd. Published 2016 by John Wiley & Sons, Ltd.

The JCT published its Tendering Practice Note 2012, which provides both general guidance and model forms for use on JCT contracts for both the pre-selection and tender phases. This is a guidance document covering the selective tendering process and the process of pre-qualification.

The principles of good practice relating to the selection of contractors are considered under the following headings:

- tender list;
- tendering procedure
  - preliminary enquiry
  - tender documents and invitation
  - tender period
  - tender compliance
  - late tenders;
- tender assessment
  - opening tenders
  - examination and adjustment of the priced document
  - negotiated reduction of tender; and
- notification of results.

Tender list

The primary aim of selection is to produce a list of contractors all of whom are capable of completing the project satisfactorily. The final choice of contractor can then be the one submitting the lowest or most economically advantageous bona fide tender.

The cost to a contractor of preparing a tender is high and is therefore reflected in the cost of building. If a large number of contractors were afforded the opportunity to tender for a project, the total amount of abortive costs would increase proportionately by the number of unsuccessful contractors submitting tenders. Whilst the total abortive costs might not directly impact the employer, the abortive costs of unsuccessful tenders are spread across projects to which successful tenders have been submitted. It therefore follows that open tendering increases the costs of projects unnecessarily and should, if at all possible, be avoided.

In order to minimise abortive tendering costs, the design team should ensure, as early in the design process as possible, that there exists a list of suitable contractors from which those invited to tender will be selected during Stage 4 of the RIBA Plan of Work 2013 (see Figure 2.1). Those employers who build frequently will normally maintain a list of suitable contractors. Such a list should be reviewed on a regular basis so that contractors who have not performed well or who have otherwise become unsuitable can be removed and those not on the list who meet the relevant criteria can be added to it. When there is no list of approved contractors, one should be compiled from firms known to the employer and the design team and/or from those who respond to press advertisements. Such advertisements need to be carefully worded and must at least indicate the size, nature and location of the project and the date when the tender documents will be ready.

It is worth noting in this case that European Union procurement rules set down contractor selection and contract award procedures applicable to public authorities in respect of contracts above certain specified values. Further information on these rules can be found within EU Procurement in Chapter 16 of this book.

Preliminary enquiry

About a month before the tender documents are due to be despatched, the selected contractors should be sent a letter containing as much of the information listed as follows as possible and asking whether they wish to tender for the project:

- project name, function and general description;
- employer;
- design team;
- location of site;
- approximate cost range;
- number of tenderers;
- form of contract and whether it is to be executed under hand or as a deed;
- any sub-contractors for major items;
- anticipated date of possession;
- contract period;
- anticipated date for despatch of tender documents;
- tender period; and
- period for which tender is to remain open.

After the latest date for response to the preliminary enquiry, the list of tenderers should be finalised and those included on the list notified. At the same time, any contractors who asked to tender by responding to an advertisement and who are not included on the tender list should be informed of this.

Tender documents and invitation

On the date named in the preliminary enquiry, the tender documents should be sent to or made available for collection by the selected tenderers. The tender documents should comprise the following:

- a checklist of all tender documents;
- instructions to tenderers;
- two copies of the drawings, schedules and specification;
- two copies of the bills of quantities or pricing schedules;
- two copies of any health and safety pre-construction information;
- two copies of the form of tender; and
- suitably addressed envelopes.

Part III

One envelope should be for the return of the tender, endorsed with the word 'Tender' and the name of the project together with the latest date and time for the tender return. Another envelope should be for the return of the priced bills of quantities or pricing schedules in support of the tender (if they are to be returned with the tender), endorsed with the project name and the words 'Priced Bills' or 'Priced Schedules' together with a space for the tenderer's name.

The instructions to tenderers referred to above should include the following:

- where, to whom and by when to submit the tender;
- any information required to be submitted with the tender, such as method statement; quality control resources and programme of work;
- method of packaging and identifying the tender;
- method for dealing with any queries relating to the tender documents;
- method of dealing with any errors or inconsistencies in the tender documents discovered after they have been issued;
- whether alternative proposals are acceptable if accompanying a compliant tender;
- notes relating to any contractor's designed portion;
- required period of validity for the tender;
- anticipated tender acceptance date;
- arrangements for inspecting additional information;
- arrangements for visiting the site;
- whether the employer will accept the lowest or any tender;
- tender assessment criteria;
- method of handling errors in tenders; and
- method of communication of tender results.

Contractors should of course acknowledge receipt of the documents and confirm that they are still able to provide a bona fide tender by the due date, or if for some reason they are unable to do so, they should return the documents immediately.

Tender period

The time required by a contractor to prepare a tender is dependent on both the size and complexity of the project. Whilst it is generally accepted that the minimum tender period should be 4 weeks, a longer period will be required in some instances, particularly so in the case of projects to be procured on a design and build basis. If realistic prices are to be tendered, it is imperative that tenderers be given sufficient time in which to: obtain competitive prices from their suppliers and sub-contractors; comply with the requirements of CDM2015; and formulate their bids properly.

Tender compliance

In order to achieve fair competitive tendering, it is essential that any unauthorised amendments to or qualifications against the tender documents by a tenderer render the tender non-compliant and subject to rejection, although the tenderer should be given the opportunity to withdraw the amendments/qualifications and stand by his tender. It is also essential that unsolicited alternative bids, in terms of either price or time, are considered non-compliant and rejected.

It is vitally important that tenderers compete on a level playing field, and this aim would be challenged by any tenderer who submits a non-compliant bid.

Late tenders

The latest time and date for the submission of tenders is best established when the tender documents are first made available. This date may, for a variety of reasons, have to be extended, in which case, all tenderers must be notified. Any tender received after the latest time and date fixed for their receipt should not be admitted to the competition and is best returned unopened to the tenderer. This is important to avoid the suggestion that a late bidder has obtained knowledge of his competitors' pricing before submitting his own bid.

Opening tenders

Tenders should be opened as soon as possible after the published time for their receipt. It is as well if at least two people are present. Upon opening, each tender should be countersigned by two of those present and the tenderer and tendered amount should be entered on a list, which, when complete, should also be signed by two of those present.

Examination and adjustment of the priced document

The lowest tenderer should be requested to provide a priced document (bills of quantities or pricing schedule) in support of the tender if this was not required to be submitted with it. The quantity surveyor should examine the priced document in support of the lowest tender to determine that any amendments notified during the tender period have been properly included and to detect any errors in computation. Any errors found are dealt with in the manner prescribed in the instructions to tenderers; it is important that the manner is laid down from the outset rather than left until the return of tenders.

The most frequently adopted methods for dealing with errors in computation are the following:

■ The tenderer is advised of the errors and is given the opportunity to confirm or withdraw his offer. If he withdraws, the examination process is repeated with the next lowest tenderer. If he is prepared to stand by his tender, the pricing document is endorsed to the effect that all rates and prices, excluding those relating to provisional sums and prime cost sums, be considered increased or decreased in the same proportion as the corrected total of the priced items exceeds or falls short of the uncorrected total of such items. It is these adjusted rates and prices that will subsequently be used for valuing variations.

- The tenderer is advised of the errors and is given the opportunity to amend his tender to correct genuine errors or to withdraw his offer. If he amends and the amended tender is no longer the lowest, or if he withdraws, the examination process is repeated with the next lowest tenderer.

When a tender appears to be free from error, or when despite error the contractor is prepared to stand by his tender, or when a tender is still lowest after amendment, such a tender should be recommended for acceptance. However, there are occasions when the tender instructions indicate that price will not be the only criterion by which the offers are to be assessed and that other factors (such as the tenderer's proposed solutions to specific problems, his proposed method of working or his proposed programme) may be taken into account, and those other factors may have weightings applied to them in order that the tenderers may formulate their bid in a manner most likely to be attractive to the employer. When factors other than price alone are to be taken into account, the design team will evaluate the bids against the relevant criteria and may conclude that the lowest price offered is not necessarily the most appropriate one to recommend for acceptance.

Negotiated reduction of a tender

Good practice dictates that a contractor's tendered price should not be changed on an unamended scope of works except for the correction of genuine errors, as noted previously. However, should the lowest tender exceed the employer's budget, it is permissible for a reduced price to be negotiated with the lowest tenderer on the basis of agreed changes in the specification and/or quantity of the work. If such negotiations fail, then, and only then, is it permissible to commence negotiations with the next lowest tenderer.

Notification of results

Immediately after the tenders have been opened, all but the lowest three tenderers should be notified that their tenders have not been successful. At the same time, the second and third lowest should be advised that they are not the lowest but that they may be approached again should the lowest tenderer withdraw his offer. As soon as a decision has been made to accept a tender, all unsuccessful tenderers should be notified and any submitted pricing documents accompanying the tender offer should be returned to them unopened.

The employer should be encouraged to decide as quickly as possible which tender to accept, for it is of the utmost importance to contractors to know whether their tenders have been successful. Once the contract has been let, it is good practice to provide each tenderer with a complete list of the tender prices received. Whilst this does not disclose who tendered which amount, it does enable each tenderer to see how they stood in relation to the rest.

Those contracts caught by the EU regulations require that unsuccessful tenderers are provided with a standstill period of at least 10 days between the date of notification of their failure and the proposed date when the contact will be entered into with the successful tenderer to allow those unsuccessful tenderers to challenge the award should they so wish (see further information on this topic in the EU Procurement Chapter 16 of this book).

Tender analysis

Frequently, the quantity surveyor will be required to analyse the successful tender, and in any event, it is good practice to do so. Traditionally, the analysis is based on the functional elements of the building (see Chapter 18) to enable the quantity surveyor to revise the cost plan to reflect the tendered prices in order to establish a proper basis for post-contract cost control (see Chapter 29) and for the purposes of adding to the industry's historical cost data bank.

More and more projects are dependent on third-party funding and government-sponsored financial incentives for their economic viability. The quantity surveyor is therefore frequently required to provide financial information, normally gleaned from tender analysis, to satisfy funders, to support grant applications and to enable the employer to take advantage of capital allowances and other forms of tax relief offered by HM Revenue and Customs.

E-Tendering

The use of documents in electronic format has become more common, and as a consequence, best practice guidance and procedures for e-tendering continue to develop and evolve along with advancing technology. The RICS guidance note on e-tendering is now in its second edition (2010) and covers such issues as the following:

- technology;
- security; and
- procedures.

Part IV

Contract Administration

Chapter 25
Placing the Contract

The placing of the contract is a relatively simple, routine matter, but the events that immediately precede it and those that immediately follow it are far from simple and are certainly of great importance.

Preparing and signing the contract documents

If the date of possession of the site by the contractor and the date for completion of the works have not previously been decided, they should be agreed with the contractor whilst the contract documents are being prepared for signing. A contract entered into without these two key markers properly identified might be regarded as baseless. At the same time, it is as well to make arrangements for the initial site meeting and to ensure that the contractor has no valid objection to any of the proposed sub-contractors and suppliers (see Chapter 23).

The contract documents, which should be so labelled, will normally comprise the articles of agreement, the conditions of contract, the drawings showing the work to be done, the priced bills of quantities (or the priced document) and any post-tender negotiation documentation. In preparing the SBC for signing, it is necessary to complete the articles of agreement and the contract particulars at the beginning together with Schedule 6 relating to the form of bonds at the end.

The information necessary for the completion of the articles of agreement and the contract particulars together with details of the above-noted deletions and amendments should have been included in the tender documents. It is imperative that no other insertions, deletions or amendments are made to the SBC without the prior agreement of the contracting parties.

The Aqua Group Guide to Procurement, Tendering and Contract Administration, Second Edition.
Edited by Mark Hackett and Gary Statham.
© 2016 John Wiley & Sons, Ltd. Published 2016 by John Wiley & Sons, Ltd.

The SBC incorporates sectional completion and contractor's designed portion (CDP) as standard text. It is worth noting in this case that sectional completion is not to be confused with partial possession by the employer. Sectional completion relates to a specific requirement on the part of the employer, recorded in the contract documents, for the works to be completed in discrete, phased sections by pre-determined dates. In contrast, partial possession by the employer is prescribed by SBC clauses 2.33 to 2.37, which provide that the employer may, with the consent of the contractor, take possession of a part or parts of the works or section at any time prior to practical completion. The same clauses also cover the associated considerations such as practical completion of the part(s) taken over, defects and insurance.

Sectional completion

Sectional completion under the SBC is for use when the works are to be completed in phased sections, of which the employer will take possession upon practical completion of each. If sectional completion is to be used, it is necessary for the works to be divided into the sections in the tender documents at tender stage. The number of sections that a project may be divided into is, in theory, infinite. However, in practical terms, the project will be influenced by physical and commercial considerations. Each section is treated as having its own date of possession and date for completion. In effect, each section is treated as being a mini-project in its own right. Should a contractor fail to complete a specific section by that section's date for completion, the contractor will be liable for the amount of liquidated damages stated in the contract particulars. The contract particulars must state the amount of liquidated damages applicable to each section.

Contractor's designed portion

The CDP is for use when the contractor is required to complete the design of a portion of the works. The use of the contractor's designed portion option creates the following additional contract documents:

- **The Employer's Requirements** – This document shows and describes the requirements of the employer in respect of that portion of the works to be designed by the contractor.
- **The Contractor's Proposals** – This document shows and describes the contractor's proposals for the design of the relevant portion of the works to meet the employer's requirements.
- **The CDP Analysis** – This document is an analysis of the portion of the contract sum relating to the portion of the works to be designed by the contractor.

It will be noted that these documents are the same as those in use with the JCT's design and build contract. Whilst the references to and obligations surrounding the CDP are spread throughout the contract, the principal clauses comprise the following:

Recitals

Ninth	the Works include the design and construction of …
Tenth	the Employer has supplied the Contractor documents … for the design and construction of the CDP
Eleventh	in response to the Employer's Requirements the Contractor has supplied the Employer
Twelfth	the Employer has examined the Contractor's Proposals and, subject to the Conditions, is satisfied that they appear to meet the Employer's Requirements.

Articles

1   Contractor's obligations

Clause

2.2	Contractor's Designed Portion
2.3	Materials, goods and workmanship
2.9.4 & 2.9.5	Construction information and Contractor's master programme

Schedule 1

2.14.4	Contract Bills and CDP Documents – errors and inadequacy
2.15.5	Notification of discrepancies etc.
2.16	Discrepancies in CDP Documents
2.17	Discrepancies from Statutory Requirements
2.19	Design liabilities and limitation
2.20	Errors and failures – other consequences
2.21.2	Fees, Royalties and Patent Rights
2.40	Contractor's Design Documents – as-built drawings
2.41	Copyright and use
3.7.2	Consent to sub-contracting
3.10.3	Compliance with instructions
3.14.3	Instructions requiring Variations
5.8	Contractor's Designed Portion – Valuation
6.12, 6.13	CDP Professional Indemnity Insurance
8.7.2.2	Consequences of termination under clauses 8.4 to 8.6
8.12.2.2	Consequences of termination under clauses 8.9 to 8.11 etc.

The contractor's main obligation under a contractor's designed portion is, as set out in clause 2.2, to complete the design portion in accordance with the contract drawings and contract bills. Depending on the level of detail contained in the employer's requirements, the contractor must develop the specification in respect of the selection of materials and goods and the standard of workmanship.

Executing the contract

The tender documents should record, at tender stage, whether the agreement is to be executed under hand or as a deed. The Limitation Act 1980 provides that

Part IV

actions for breach of contract can be commenced within 6 years of the breach in respect of a contract executed under hand and within 12 years of the breach in respect of a contract executed as a deed.

The attestation clauses in the SBC make provision for the agreement to be executed under hand or as a deed both by a company or other body corporate and by an individual. Irrespective of how the agreement is executed, both parties to the contract should initial each and every contract document. Where applicable, the parties should also initial all amendments made to SBC (or, for that matter, any standard form that is the subject of amendment).

Performance bonds and parent company guarantees

A performance bond is a tripartite agreement between the contractor, a surety and the employer that provides for the surety to pay a sum of money to the employer in the event of default by the contractor. The value of the bond normally provided is equal to 10% of the original contract sum. As the raising of a bond invariably imposes a financial burden on the contractor in terms of borrowing, it is considered to be good practice for the bond to be released coincidentally with the achievement of a particular event, for example, practical completion or the issue of the certificate of making good.

The contractor will normally obtain such a bond from an insurance company or a bank. However, where the contractor is a subsidiary of a large organisation, it may well be that a guarantee from the parent company will suffice instead of a bond. The bond holder and the terms of the bond or guarantee must be approved by the employer. A typical performance bond (Example 25.1) and a typical parent company guarantee (Example 25.2) are reproduced on the following pages. The JCT SBC includes standard bonds in its Schedule 6 for advance payment, payment for off-site materials and goods and retention, but no performance bond. Whether a bond or guarantee is required must be determined prior to tender stage and the appropriate entry made in the tender documents so that the tenderers can make due allowance for all associated costs in their tendered prices.

Collateral warranties

Collateral warranties in general and those to be provided by consultants in particular are considered in Chapter 2, where mention is made of the standard warranty agreements published by BPF and CIC.

Similarly to consultants, main contractors are also frequently required to enter into collateral warranty agreements and are always required to do so when those entered into by the consultants are the BPF agreements CoWa/F and CoWa/P&T or the CIC agreements CIC/ConsWa/F and CIC/ConsWa/P&T, under which the responsibilities of the consultants are on the basis that the main contractor will provide contractual undertakings to the funding institution, purchaser or tenants, which are of similar effect to those provided by the consultants. Section 7 of the

THIS BOND is made the.........................day of................... 20........

BETWEEN ....

of ....

(hereinafter called 'the Contractor') of the first part.

and ....

of ....

(hereinafter called 'the Surety') of the second part.

and ....

of ....

(hereinafter called 'the Employer') of the third part.

1. By a Contract dated.......... made between the Contractor and the Employer (hereinafter called 'the Contract') the Contractor has agreed to carry out the Works specified in the Contract for the sum of £ ....

2. The Contractor and the Surety are hereby jointly and severally bound to the Employer in the sum of £.........................(not exceeding ten per cent of the original Contract Sum) which sum shall be reduced by an amount equal to ten per cent of the value of any part or parts of the Works taken into possession of the Employer under the provisions of the Contract *(or of any Section of the Works upon the Architect/Contract Administrator certifying practical completion of that Section) provided that if the Contractor shall, subject to Clause 5 hereof, duly perform and observe all the terms, conditions, stipulations and provisions contained or referred to in the Contract which are to be performed or observed by the Contractor or if on default by the Contractor the Surety shall, subject to Clause 3 hereof, satisfy and discharge the damage sustained by the Employer thereby up to the amount of this Bond then this agreement shall be of no effect.

* The wording in brackets should be deleted unless Sectional Completion applies.

3. If the Contractor has failed to carry out the obligations referred to in Clause 2 hereof then written notice requiring the Contractor to remedy his failure, where possible, shall be given, and if the Contractor fails so to do or repeats his default or if the Contractor's employment under the Contract is determined in accordance with clause 8.4 of that Contract the Employer shall be entitled to call upon the Surety in accordance with Clause 2.

Example 25.1   Performance bond.

4. Any alteration to the terms of the Main Contract or any variations required under the Contract shall not in any way release the Surety from its obligations under this agreement.

5. The Contractor and the Surety shall be released from their respective liabilities under this agreement upon the date of the Practical Completion of the Works as certified by the Architect/Contract Administrator appointed under the Contract.

IN WITNESS whereof the parties hereto have executed this Document as a Deed the day and year first before written.

Contractor: ....

Signed by (insert name of Director) and   Director

(insert name of Company Secretary or   ....

second Director) for and on behalf of   Director/Company Secretary

....

Surety: ....

Signed by (insert name of Director) and   Director

(insert name of Company Secretary or   ....

second Director) for and on behalf of   Director/Company Secretary

....

Employer: ....

Signed by (insert name of Director) and   Director

(insert name of Company Secretary or   ....

second Director) for and on behalf of   Director/Company Secretary

....

Example 25.1   (Cont.)

SBC contains provision for JCT warranties to be entered into and refers to part 2 of the contract particulars. JCT warranties comprise the following:

- CWa/F – Contractor Collateral Warranty for a Funder; and
- CWa/P&T – Contractor Collateral Warranty for a Purchaser or Tenant.

The JCT also publishes standard subcontractor warranties for use with the SBC for the funder (SCWa/F), purchaser and tenant (SCWa/P&T), and the employer (SCWa/E).

THIS AGREEMENT is made the.......................day of.................. 20.......

BETWEEN ....

whose registered office is at ....

(hereinafter called 'the Guarantor') of the one part and ....

whose registered office is at ....

(hereinafter called 'the Employer') of the other part,

WHEREAS

A   ....

        (hereinafter called ....

        a subsidiary of the Guarantor has entered into a Contract with the Employer of even date to

        carry out certain works at ....

        (hereinafter called 'the Contract')

B.   The Guarantor has agreed to guarantee the due performance by ....

        ....of the Contract in the manner hereinafter appearing:

NOW THE GUARANTOR HEREBY AGREES WITH THE EMPLOYER as follows:

1.   If..................(unless relieved from the performance by any clause of the Contract or by statute or by the decision of a tribunal of competent jurisdiction) shall in any respect commit any breach of its obligations or duties under the Contract then the Guarantor will procure the remedying of any breach of..................obligations failing which the Guarantor will pay to the Employer and/or his assigns the amount (subject as hereinafter provided) of all losses, damages, costs and expenses which may be incurred by the Employer by reason of any default on the part of..................in performing its obligations and duties or otherwise to observe and comply with the provisions contained in the Contract to the extent that such losses, damages, costs and expenses are or would otherwise be recoverable by the Employer from.....................under the said Contract.

2.   The Guarantor further agrees with the Employer that it shall not in any way be released from liability hereunder by any alteration in the terms of the Contract made by agreement between the Employer and.....................or in the extent or nature of the Works to be constructed, completed and maintained thereunder or by any allowance of

Example 25.2   Parent company guarantee.

time or forbearance or forgiveness in or in respect of any matter or thing concerning the Contractor or by any other matter or thing whereby (in the absence of this present provision) the Guarantor would or might be released from liability hereunder, and for the purposes of this Agreement any such alteration shall be deemed to have been made with the consent of the Guarantor and Clause 1 hereof shall apply in respect of the Contract so altered. Provided always that the amount of the Guarantor's liability under this Agreement shall be no greater than the amount which would have been recoverable against the Contractor by the Employer under the Contract and the same limitation periods fixed by statute which apply to the Contract shall equally apply to this Agreement.

IN WITNESS whereof the parties hereto have executed this Document as a Deed the day and year first before written.

Guarantor: ....

Signed by (insert name of Director) and   Director

(insert name of Company Secretary or   ....

second Director) for and on behalf of   Director/Company Secretary

....

Employer: ....

Signed by (insert name of Director) and   Director

(insert name of Company Secretary or   ....

second Director) for and on behalf of   Director/Company Secretary

....

Example 25.2   *(Cont.)*

Third party rights

The Contracts (Rights of Third Parties) Act 1999 referred to in Chapter 2 affords parties to a contract a viable alternative to collateral warranties. The SBC expressly affords parties to the contract the choice of either third party rights under the Act or collateral warranties, should the need to circumvent the doctrine of privity of contract be required. The SBC envisages that those requiring such rights are those who would ordinarily require collateral warranties, for example, funders, purchasers or tenants. Section 7 of the SBC covering third party rights requires that, in respect of purchasers or tenants, the class or description of those with the benefit of the rights is made known from the outset and that the contract particulars state for which part of the works such rights are required.

Issue of documents

The SBC requires that the contract documents remain in the custody of the employer so as to be available at reasonable times for inspection by the contractor. It also requires that immediately after the contract has been executed, the architect/contract administrator must provide the contractor with:

- one copy of the contract documents certified on behalf of the employer;
- two further copies of the contract drawings; and
- two copies of the unpriced bills of quantities or specification/schedules of work.

The SBC also requires that as soon as possible after the execution of the contract:

- the architect/contract administrator shall provide the contractor with two copies of any descriptive schedules or other similar documents necessary for use in carrying out the works together with any pre-construction information required for the purposes of regulation 4 of the CDM Regulations;
- the contractor shall provide the architect/contract administrator with his master programme for the execution of the works identifying, where required in the Contract Particulars, the critical paths and/or providing such other details as are specified in the Contract Documents;
- the contractor shall provide the architect/contract administrator with copies of the contractor's design documents and are related calculations as are reasonably necessary to explain or amplify the contractor's proposals;
- the contractor shall provide the architect/contract administrator with copies of all the levels setting out dimensions, which the contractor has prepared for completing any CDP.

The contractor's design documents are required to be provided as and when necessary in accordance with the contractor's design submission procedure and clause 2.9.5 and Schedule 1 of the SBC.

Insofar as they are relevant to the project, the following documents will also have to be provided to the contractor by the design team:

- party wall agreements;
- condition surveys of adjoining properties;
- conditional planning approvals;
- tree preservation orders;
- building control approval and notices to be served during the course of the works; and
- procedures and estimates for works to be carried out by statutory and local authorities.

The contractor is required by the SBC to keep one copy of each of the following documents on site so as to be available to the architect:

- contract drawings;
- the unpriced bills of quantities or specification/schedules of work;

Part IV

- CDP documents;
- descriptive schedules or other similar documents;
- the master programme; and
- any further drawings or details issued to explain and amplify the contract drawings.

Insurances

The insurance provisions in the SBC are to be found in Section 6 and Schedule 3.

Clause 6.4.1 requires the contractor to take out and maintain insurance against claims for personal injury to or death of any person or for loss, injury or damage to real or personal property that arises out of or in the course of or is caused by the carrying out of the works except for claims caused by a default of the employer or those for whom the employer is responsible. The insurance in respect of the contractor's liability to third parties must also extend to meet any like claims made against the employer by third parties. Insurance in respect of injury or death to an employee of the contractor must comply with the current legislation (at the time of writing this is the Employers' Liability (Compulsory Insurance) Regulations 1998 and with Amendment Regulations 2008). Insurance in respect of all other claims has to be for no less cover than is stated in the contract particulars.

Clause 6.5.1 also provides the employer with the option to require the contractor, by way of an architect's instruction, to insure (in the names of the employer and the contractor) against any expense, liability, loss, claim or proceedings incurred by the employer as a result of injury or damage to any property, other than the works and any site materials for the works, caused by:

- collapse;
- subsidence;
- heave;
- vibration;
- weakening or removal of support; and/or
- lowering of ground water.

arising out of or in the course of or by reason of the carrying out of the works.

The aforementioned exclusions would not apply to injury or damage:

- caused by the default of the contractor;
- caused by errors or omissions in designing the works;
- which can reasonably be foreseen to be inevitable with regard to the nature of the work;
- which it is the responsibility of the employer to insure under the SBC Option C (if applicable – see the following section);
- arising from the consequence of war and like acts;
- caused by the excepted risks (as defined in SBC clause 6.8);
- caused by pollution or contamination; and
- resulting in damages payable by the employer for breach of contract.

The extent of cover required has to be stated in the contract particulars, but as this particular insurance is instigated by an architect's instruction, the cost to the contractor of taking it out and maintaining it is added to the contract sum as an extra.

Clause 6.7 deals with insurance of the works and Schedule 3 provides three alternatives as follows:

- **Option A** – where the works comprise new buildings and the parties agree that the contractor is required to take out and maintain a joint names policy for all risks insurance of the works.
- **Option B** – where the works comprise new buildings and the parties agree that the employer is required to take out and maintain a joint names policy for all risks insurance of the works.
- **Option C** – where the works are in or are extensions to existing structures. In this case, the employer is required to take out and maintain a joint names policy for loss or damage caused to the existing structures, the Works thereto and contents by one or more of the specified perils and a joint names policy for all risks insurance of the works.

The term 'joint names policy' is expressly defined in the SBC in clause 6.8 as being 'a policy which includes the Employer and the Contractor as composite insured and under which the insurers have no right of recourse against any person named as an insured, or, pursuant to clause 6.9, recognised as an insured thereunder.'

Clause 6.9 further provides that the joint names policy must either provide for recognition of each sub-contractor as an insured or include a waiver by the insurers of any right of subrogation that they may have against any such sub-contractor. Subrogation refers to the process by which an insurer, who has paid the insured for loss or damage suffered, is entitled to sue, in the insured's name, whoever caused the loss or damage.

The amount of cover to be provided by the joint names policy is the full reinstatement value of the works plus the percentage stated in the contract particulars to cover professional fees and, in the case of works in or extensions to existing structures, the full value of the existing structures and of their contents owned by the employer or for which the employer is responsible.

Clause 6.7 of the SBC under Option A.3 recognises that it is common practice for contractors to maintain annually renewable insurance policies that provide appropriate Option A all risks insurance and sets out 'deemed-to-satisfy' provisions in respect of such annual policies.

The SBC, amongst other things, defines terrorism and terrorism cover and provides the employer with the option of determining the employment of the contractor or of requiring the contractor to complete the works at the employer's expense if insurers withdraw terrorism cover during the currency of the contract. These provisions were previously contained in JCT Amendment TC/94, issued April 1994. Following terrorist attacks on the UK mainland, particularly the IRA bombings in the City of London, which caused billions of pounds worth of damage, insurance against terrorism had become unavailable. In order to comply with the SBC requirement to maintain cover, the government agreed to act as 'reinsurer of last resort'.

Construction insurance is an extremely complex subject that is not helped by the fact that available policies do not incorporate standard wording. Consequently, the employer should always be advised to consult a specialist insurance adviser. Footnotes to the SBC insurance clauses remind users that it is sometimes not possible to obtain insurance in the precise terms required by SBC, in which case the conditions must be amended accordingly. Once insurances are in place, it is important that they are maintained and renewed when necessary. It is therefore recommended that the design team checks that insurances are maintained and renewal premiums are paid when due. It is advisable to keep copies of premium receipts or brokers' confirmatory letters on file.

Chapter 26
Meetings

Initial meeting

As soon as practicable after the contract has been placed, the building team should meet. Although this initial meeting may take place on site, it will more probably take place in the offices of the project manager, the architect/contract administrator or the main contractor.

The manner in which the initial meeting is conducted will greatly influence the success of the project, and succinct clear direction from the chair will hopefully provide a strong inducement to a similar response from others. Since at this stage the person having the most complete picture of the project is likely to be the project manager (if there is one) or the architect/contract administrator (if there is not), it is logical that the project manager or the architect/contract administrator should take the chair. The arrangements for this meeting should be discussed with the main contractor beforehand. It is suggested that as many of the following as are involved with the project should be requested to attend:

- employer;
- project manager;
- principle designer;
- architect/contract administrator;
- quantity surveyor;
- structural engineer;
- building services engineer(s);
- contractor; and
- clerk of works.

The Aqua Group Guide to Procurement, Tendering and Contract Administration, Second Edition.
Edited by Mark Hackett and Gary Statham.
© 2016 John Wiley & Sons, Ltd. Published 2016 by John Wiley & Sons, Ltd.

It is also suggested that the agenda for the initial meeting should include the following matters:

- introductions;
- factors affecting the carrying out of the works;
- programme;
- sub-contractors and suppliers;
- lines of communication;
- bonds, collateral warranties and insurances (see Chapter 25);
- financial matters; and
- procedure to be followed at subsequent meetings.

Introductions

The introduction of those attending needs no elaboration, although it is more than just a formality as it establishes the initial contact between individuals who must work together in harmony if the project is to run smoothly.

Factors affecting the carrying out of the works

These should be described fully in the contract documents but may require emphasis and clarification at the initial meeting. Such factors may include but are not limited to the following:

- existing mains/services;
- site investigation, including soils and ground water information;
- access to the site;
- parking restrictions;
- use of the site;
- risks to health and safety;
- health and safety plans;
- work outside the site boundary;
- concurrent work by others;
- named and/or listed sub-contractors;
- equivalent products;
- Considerate Constructors Scheme;
- photographic records;
- prohibited products;
- protection of products;
- samples;
- building lines, setting out, critical dimensions and tolerances;
- co-ordination of engineering services;
- quality control;
- use of explosives and pesticides and on-site burning of rubbish;
- fire prevention;

- protection of the works, existing retained features, adjoining buildings, work people and the public;
- working hours and overtime working;
- location of temporary accommodation, spoil heaps and so on;
- temporary fences, hoardings, screens, roads, name boards and so on; and
- temporary services and facilities.

Programme

Clause 2.9.1.2 of SBC requires that the contractor shall provide the architect/contract administrator with his master programme for the execution of the works as soon as possible after the execution of the contract.

In addition, the contractor is also required to update it within 14 days of any decision by the architect/contract administrator that creates a new completion date. In any event, the master programme should be reviewed regularly and updated whenever necessary. It is to be noted that the master programme is not a contract document and SBC clause 2.9.3 makes it clear that nothing contained in the master programme can impose any obligation on the contractor beyond those obligations imposed by the contract documents. Some forms of contract do not require the contractor to provide a programme. In such circumstances, if a programme is wanted, the need should be stated in the bills of quantities, specification or the contract documents themselves.

In *Glenlion Construction Limited -v- The Guinness Trust* it was decided that:

- the contractor is entitled to complete the works on or before the date for completion stated in the contract particulars, or any later completion date fixed pursuant to the appropriate conditions of contract; and
- the contractor is not entitled to expect the design team to provide design information to enable him to complete the works before the date for completion stated in the contract particulars, or any later completion date fixed pursuant to the appropriate conditions of contract.

At first sight, these two statements may appear to conflict with each other, but with further thought, it will be apparent that they do not. This is because even if the contractor is capable of completing the works before the date for completion, it may not be possible for the design team to provide design information sooner than would be necessary to meet the contract completion date.

Prior to the initial meeting, the contractor should determine, as far as possible, major material delivery dates and required construction periods from sub-contractors. From this information and the Information Release Schedule, when provided by the employer (see SBC fifth recital and clause 2.11), the contractor should formulate a draft master programme for the works, which should then be circulated to all members of the building team. Whether or not an information release schedule has been provided, it is quite usual for the contractor to indicate on the draft master programme the latest dates by which drawings, schedules, instructions for placing orders and other information are required from the design team.

Part IV

The draft master programme should be considered and, if necessary, adjustments agreed during the initial meeting. Following this, the contractor should prepare his definitive master programme and circulate it to all concerned. A simple bar chart programme is given as Example 26.1 at the end of this chapter.

The importance of adhering to dates once agreed, whether they be in respect of materials deliveries, execution of the work or the production of further necessary information by the design team, cannot be over-stressed.

Sub-contractors and suppliers

Many contractual problems arise as a result of the contractor failing to issue instructions in relation to sub-contractors and suppliers timeously and as a result of poor communications between the main contractor and those firms engaged. It is therefore advisable at the initial meeting for the architect/contract administrator to inform the main contractor of his intended instructions that might affect sub-contractors and suppliers. It is also advisable to make it clear to the main contractor that the sub-contractors and suppliers are contractually his responsibility.

It is essential, for the smooth and efficient running of the project, that all of the sub-contractors are engaged sufficiently early for the work concerned to be integrated into the contractor's programme of work without causing disruption. It is also essential that all members of the building team allow sufficient time in their work schedule to carry out properly the various complex and often time-consuming procedures involved in the selection and appointment of sub-contractors and suppliers. Following the correct procedures and providing the appropriate documentation when appointing sub-contractors is of the utmost importance and is dealt with in Chapter 23.

Lines of communication

It is important at the initial meeting for the chairman to emphasise the need for all members of the building team to abide by the formal lines of communication provided for in the SBC and to make reference to the golden rules of communication given in Chapter 1.

Financial matters

It is desirable at the initial meeting for the quantity surveyor to run through the various procedures set down in the SBC relating to financial matters. The sheer number of clauses relating to financial matters might mean that the quantity surveyor has to limit the discussion on procedures to those regarding interim valuations, valuation of variations and payment. This can, however, be tailored to the employer and the contractor's experience and the specifics of the project. Clauses concerning financial matters and which might require a degree of explanation comprise the following:

2.21   Fees or charges legally demandable
2.22   Royalties and patent rights – Contractor's indemnity

Part IV

6.13   CDP Professional Indemnity Insurance – Increased cost and non-availability

8.4   Termination by Employer – Default by Contractor

8.5   Termination by Employer – Insolvency of Contractor

8.6   Termination by Employer – Corruption

8.7   Termination by Employer – Consequences of termination under Clauses 8.4 to 8.6

8.8   Termination by Employer – Employer's decision not to complete the Works

8.9   Termination by Contractor – Default by Employer

8.10   Termination by Contractor – Insolvency of Employer

8.11   Termination by either Party

8.12   Consequences of termination under Clauses 8.9 to 8.11, etc.

Schedule 2 – Valuation and Acceleration Quotation procedures;
Schedule 6 – Forms of Bonds;
Schedule 7 – Fluctuations options; and
Schedule 8 – Supplementary provisions.

It is worth stressing that no adjustment in respect of varied work will be made to the contract sum unless the matter is covered by an architect's instruction issued in accordance with the terms of the contract. Also, it is worth reminding the contractor of any other financial matters required by the contract documents, such as the provision by him of a cash flow forecast showing the gross valuation of the works at the date of each interim certificate.

Procedure to be followed at subsequent meetings

Formal site meetings are not to be confused with the architect's periodic site visits and the numerous meetings between the main contractor and others that are necessary for the progress of the works. The frequency of formal site meetings will vary with the size and complexity of the project, the stage of the job and any difficulties being encountered. It would be unusual for such meetings to be held less frequently than once every 4 weeks.

The requirements of the employer as to the nature, normal frequency, procedures and location for formal site meetings should be given in the contract documents. These meetings will be chaired by the project manager (if there is one) or the architect/contract administrator (if there is not) and will normally be attended by the design team, the main contractor and such sub-contractors as are requested to be present. The object of the formal site meetings is to provide a forum for the presentation by the contractor of his progress report, which should include at least:

- a progress statement by reference to the master programme for the works;
- details of any matters materially affecting the regular progress of the works; and
- any requirements for further drawings, details and instructions.

It also provides an opportunity for contractors to raise any queries they may have with the design team as a whole and to report formally on health and safety issues.

The project manager or architect/contract administrator should notify the rest of the design team and the main contractor of the dates and times for site meetings and should request those whose presence is required to attend. It should be the responsibility of the main contractor to invite to the site meetings representatives of those sub-contractors and suppliers whom he or the design team wishes to be present. Due consideration should be given to the value of people's time, with care being taken not to invite to meetings those whose presence is not really necessary.

The project manager or architect/contract administrator and the main contractor should agree the agenda for each site meeting in advance. A standard form of agenda is useful as a model and an *aide memoire*. A typical site meeting agenda is given as Example 26.2 at the end of this chapter. The minutes of site meetings will normally be taken and distributed by the chairman of the meeting. They must be impartial and concise and accurately record all decisions reached and actions required. Typical site meeting minutes are given as Example 26.3.

It is worthy of note that a badly run site meeting can do untold harm and be a serious waste of time for all attending, whereas a well-managed meeting can be a great aid to the smooth running of the project. Furthermore, there is little doubt that well-run meetings maintain the team's interest in the project and impart a sense of involvement and urgency, which is difficult to achieve by any other means.

Contractor's meetings

The main contractor will normally hold meetings with sub-contractors and suppliers shortly before the formal site meetings. Such meetings help facilitate the accurate reporting of progress and ensure that all sub-contractors and suppliers have their information requirements noted and met. They also provide an opportunity to establish requirements for holes, chases, recesses, fixings and the like before work is put in hand and thereby avoid conflict with other work.

Employer's meetings

The employer may also hold separate meetings with the rest of the design team during the progress of the works. Such meetings are normally used to discuss general progress and to resolve any outstanding design issues, such as colour schemes. They also provide an opportunity to consider the possible effects of proposed variations on the works and on the anticipated final costs.

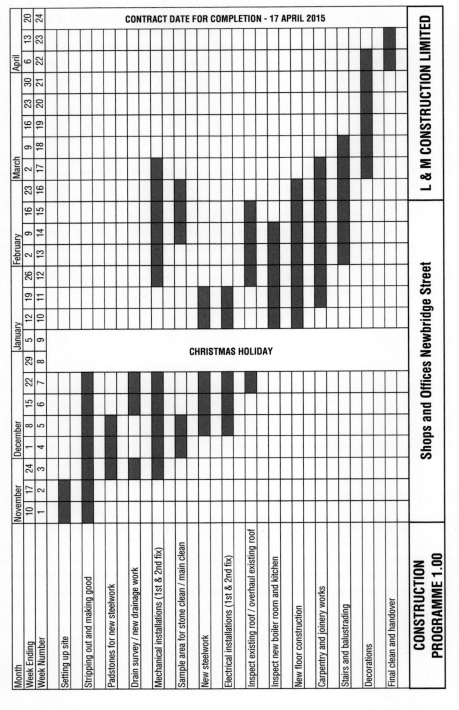

Example 26.1 Construction programme.

Project:	Shops & Offices, Newbridge Street, Borchester		
Project ref:	456		
AGENDA FOR SITE MEETING			
Date:	17 November 2014 at 10.00am		
1.0	Apologies		
2.0	Minutes of last meeting		
3.0	Contractor's report General report		
	Progress statement by reference to the master programme for the works		
	Details of any matters materially affecting regular progress of the works		
	Information received since last meeting		
	Requirements for further drawings, details and instructions		
	Health and safety matters		
4.0	Clerk of Works' report		
	Site matters		
	Quality control		
	Lost time		
5.0	Design Consultants' reports		
	Architect		
	Structural Engineer		
	Building Services Engineer		
6.0	Quantity Surveyor's report		
7.0	Contract completion date		
	Assess likely delays		
	Review factors from previous meeting		
	List factors for review at next meeting		
	Record anticipated completion date		
8.0	Any other business		
9.0	Date and time of future meetings		
	Site meetings		
	Site visits		

Distribution:

Copies	2 Employer	3 Main Contractor	1 Project Manager
	1 CDM Co-ordinator	1 Architect	1 Quantity Surveyor
	1 Structural Engineer	1 Services Engineer	1 Clerk of Works

Example 26.2 Typical site meeting agenda.

Project:	Shops & Offices, Newbridge Street, Borchester	Ivor Barch
Project ref:	456	Associates
		Prospects Drive
		Fairbridge

SITE MEETING NO 4

Date:	17 December 2014	
Location:	Site	
Present:	S Gilbert	Employer (OWGS Ltd)
	A Morley	Main Contractor (L & M Construction Ltd)
	G Mackay	Main Contractor (L & M Construction Ltd)
	B Hunt	Project Manager
	Ivor Barch	Architect (Ivor Barch Associates)
	R W Pipe	Quantity Surveyor (Fussedon Knowles & Partners)
	I Tegan	Structural Engineer (GFP & Partners)
	F Adams	Services Engineer (Black & Associates)
	H Hemmings	Clerk of Works

Item		Action
1.0	Apologies	
	None.	
2.0	Minutes of last meeting	
	Agreed as correct.	
3.0	Contractor's report	
3.1	Progress is generally satisfactory.	
3.2	Still one week behind master programme due to late delivery of bricks.	
3.3	AI5 received and actioned.	
3.4	Details of ironmongery revisions required in next two weeks.	
3.5	No accidents to report.	Architect
4.0	Clerk of Works' report	

Example 26.3   Typical site meeting minutes.

4.1	Concern expressed about poor stacking of bricks.	L&MC
5.00–8.00	Continue through agenda	
9.0	Future meetings	
	Site Meeting- 4 January 2014 at 10.00am.	All
	Architect's site visit - 21 December 2014 at 10.00am.	

Distribution:

Copies	2 Employer	3 Main Contractor	1 Project Manager
	1 CDM Co-ordinator	1 Architect	1 Quantity Surveyor
	1 Structural Engineer	1 Services Engineer	1 Clerk of Works

Example 26.3   *(Cont.)*

Chapter 27
Site Duties

Peter Ullathorne

The architect on site

The development of drawings, models and documentation made it possible in the post-medieval world for the designers of buildings to be able to leave the site environment, to be based in a studio and have a more distant relationship with their clients, their creations – and their builders. This meant that the architect became an independent entity and lost his dual role as architect and master builder combined.

The result was that from time to time, depending on the accuracy of the architects' drawings and the skill of the builders, the designers had to visit their sites to check on the progress of the works and provide interpretation, guidance and leadership as the job proceeded. Clients looked to the architect, not the builder, as their representative and the person responsible for successful outcomes. Louis Khan defined architecture as the 'thoughtful making of spaces.' Architects are the people who hold the overall picture of the project in their minds, which they have created as a translation of their clients' instructions. Naturally, they want to see their thoughts made real with efficiency and competence.

The architect on site has the usual duty to exercise reasonable skill, care and diligence in accordance with the normal standards of the Architect's profession.

Professional services are now provided in an age of litigation. Clients endeavour on an unprecedented scale, to pass project risks to their consultants and contractors. The amount of Professional Indemnity insurance held by an architect is checked by potential clients in as much detail as his creative and technical skills. It is not surprising, in this brittle and legally constrained climate, that the definition of site visits by architects has to be worded precisely so as not to imply a depth and rigour of scrutiny that would be either uneconomic or unfeasible for them

The Aqua Group Guide to Procurement, Tendering and Contract Administration, Second Edition.
Edited by Mark Hackett and Gary Statham.
© 2016 John Wiley & Sons, Ltd. Published 2016 by John Wiley & Sons, Ltd.

to provide and which would open them to legal action for errors or omissions beyond their control or reasonable level of responsibility.

An architect's site duties vary from those that apply during the feasibility and design stages, which may comprise both subjective and objective observations, complying with the requirements of the project, including making a detailed site survey, to the highly formal site visits as prescribed by the form of contract between client and contractor.

Every participant in the sponsorship, design, project management and construction process tries hard to place risk in the hands of others, but risk has to rest somewhere. Architects have had to describe their on-site services with accuracy to avoid being burdened with more unnecessary and unwelcome responsibilities than is reasonable as these carry expensive risk.

The RIBA Standard Agreement for the appointment of an Architect (S-Con-07-A) dated 2007 and subsequent revisions – (currently RIBA Standard Agreement 2010 (2012 Revision): Architect, says that the architect should make visits to the Works in connection with the architect's design. The Agreement will stipulate the number of visits to site; however, this is unrealistic as the number of necessary site visits cannot be predicted. Site visits are rarely short or convenient. Should more site visits be considered necessary by the architect and the rest of the design team (e.g. in works to an existing building), then the client should be asked for fees commensurate with the number of visits that the demands of the project require. In SFA 2010 (2012 revision), architects are now under a contractual duty (along with other duties) to inform their clients of any issue that may affect the quality of the project. Architects are more likely to collaborate rather than co-ordinate information from others, and this revised duty to collaborate will extend to inspection and supervision activities on site. Case law dictates that architects will make regular visits to the site and any other visits made necessary by issues that have to be dealt with.

The definition of 'site' includes places where components are fabricated off site.

The RIBA Plan of Work 2013 in Work Stage 5 includes the requirement for an 'Administration of Building Contract, including regular site inspections and review of progress' and 'resolution of Design Queries from site as they arise'. Architects may visit site in their duties as architect or as the contract administrator, provided their appointment and the form of contract used between the client (employer) and the contractor provide for this role.

Architects visit sites for these main reasons:

- to understand the opportunities and constraints of the site in its setting;
  - to measure the site;
  - to hand-over the site to the contractor as part of RIBA Stage 5 (Construction), ensuring that necessary instructions are given and seeing evidence that the obligatory insurances are in place;
  - to supervise and inspect the works on site, checking that the quality of completed work conforms to agreed quality standards;
  - to assess progress to issue fair certificates;
  - to trouble-shoot, mitigating delay and cost and improving buildability, and to register defective work and require satisfactory remedies;
- to check that the contractor's progress conforms to the programme; and

Part IV

■ to check that designs are being completed according to plans and designs, within statutory consents and in compliance with regulations and standards.

The architect's duty of inspection and supervision

Case law has defined the architect's duty of inspection and supervision. These duties may start before a project starts on site, for example, where elements of the project are produced off site. Two examples of this would be prefabricated bathroom units or cladding panels. The architect has a duty to inspect and supervise periodically. The contractor also has a duty of supervision. But it is essential that practitioners achieve an understanding of the case-law definition of 'periodically'. The amount of fee must allow for all necessary inspections to be made. The frequency and duration of onsite inspections are related to the nature of the job, not the timing of, for example, regular site meetings. The architect is allowed to instruct the contractor not to cover up works that must be inspected – and it's no excuse for an architect accused of failure to carry out proper inspections that the contractor did cover up work that should have been seen beforehand! However, the architect is only required to make a reasonable examination of the works and is not generally required to pay attention to every detailed matter. Convention acknowledges that not every defect will be spotted. No architect warrants that his inspection will reveal or prevent defective work and that an architect's performance cannot be judged by the results achieved. The contractor must shoulder his proportion of the blame in that situation. The architect must prepare records of inspections identifying defects and remedies and inspections must be carried out by an architect with sufficient authority and experience within his practice – for example, a partner or associate. Interim certificates will take account of defects identified at inspection such that they can be remedied in tandem with the progression of the outstanding works, culminating in Practical Completion by which time the only thing outstanding should be some 'snagging' of minor outstanding matters.

Supervision and Inspection duties

It is important to distinguish between supervision and inspection duties.

Supervision is the responsibility of the main contractor who is generally bound by the conditions of the Building Contract to carry out and complete the works in accordance with the Contract Documents including his compliance with many criteria including time, quality and cost. The contractor controls his resources, not the architect. However, this does not excuse the architect from having a duty to supervise the general course of the work. Other forms of contract may also provide for or imply, contractor supervision, for example, a design and build arrangement.

Architects, under the terms of their employment, will normally be required to visit the site, at intervals appropriate to the stage of construction and the contractor's performance, to inspect the progress and quality of the works and to determine that they are being executed in accordance with the contract documents. While on site, it is often possible for them to make sure that problems, which are bound to arise even on the smallest project, are satisfactorily resolved. However, architects are not normally required to make frequent or constant inspections unless the situation demands this as case law stipulates. The term 'inspection' has far-reaching implications, implying a depth and rigour of examination to which many architects feel uncomfortable, and for which they might be open to legal recourse should any problems arise, which arguably should have been discovered during 'inspections'. The nature and extent of supervision arrangements will depend on the size and complexity of the works. On a small project, for example, the contractor may be able to rely on a competent general foreman, while on large projects, if the job is to be properly supervised, a number of foremen and assistants may be required, all working under a site agent or a contracts manager.

Similarly, the architect on a small project may be able to undertake his normal duties by periodic visits; however, on large and complex projects, one or more clerks of works may be needed, with possibly a resident architect. In this chapter, it is assumed that there will be a clerk of works acting with the authority of clause 3.4 of SBC/XQ, 'as inspector on behalf of the employer under the directions of the architect/contract administrator'. The role of Clerk of Works may vary slightly depending on whether the appointment is made by the employer or the architect. If appointed by the employer, the role may approach that of a project manager, looking after the employer's broader interests.

Routine site visits

Routine site visits take place in RIBA Stage 5, once the contract has been let and the architect is empowered to administer the contract. In some cases, the architect will have an inspection and reporting role only, for example, where the design team has been novated to the contractor and the client retains the services of an architect to report to him directly.

The architect will normally have more time to observe the work when making routine site visits dictated by the nature of the project, than on those occasions when a formal site meeting is held. The frequency of such visits and the depth of observation will depend on the size and complexity of the work, the speed of the progress being made and any complications arising as the work proceeds. The need for valuations may add to the frequency of site visits. Generally, site visits at fortnightly intervals and valuations at 1 month intervals are appropriate for most jobs.

Architects and their representatives are given right of access to sites under most forms of contract, as well as places where work is being fabricated by contractors off-site.

As a matter of safety and courtesy, the architect should make his presence on site known, preferably by prior arrangement, to the clerk of works and the site agent. The latter will normally accompany the visiting architect around the job. The architect should never give instructions to a workman directly. The site agent or general foreman employed by the main contractor is usually the only person designated as 'the competent person-in-charge' under the Contract to receive and act upon architect's instructions. Any oral instructions given during a site visit should be confirmed in writing (see Chapter 28).

Part IV

When making a site visit to observe the works, it is easy to be distracted and to overlook items that require attention. It is a good plan, therefore, for the architect to list in advance particular points to be looked at and any special reason for doing so. For this purpose, a standard checklist is a helpful *aide memoire*. Provided that the contents of such a list are of a general nature, they can then be amplified or adapted for each job according to the type of work and form of construction involved. A typical standard checklist is given as Example 27.1 at the end of this chapter.

It is important that accurate and full records are kept of all site visits, together with the results of any tests ordered or completed. One of the key reasons an architect visits a site is to observe and check the quality of the contractor's work against the agreed method statement. Carefully structured reports should contain site photographs, notes and sketches.

Consultants' site visits

In addition to the architect, the structural engineer and services consultants will also make regular site visits to observe the works.

Depending on the nature of the work, it is often necessary, in the early stages of a project, for the structural engineer to confirm or adjust foundation designs following excavation and to advise on works necessary to stabilise existing buildings. Subsequently, the structural engineer's main concern is in the testing of concrete and other structural elements (see *Samples and testing* in the following).

Once the services installations are under way, the services consultants are likely to undertake site visits to ensure that the works are meeting the specified requirements.

Inspections by statutory officials

Visits by statutory officials, or private sector professionals with delegated powers, are to be anticipated on any project – with a wider range visiting the larger projects.

Foremost is the building control officer from the local authority, or an approved inspector under the Building Act 1984. In addition, under the Competent Person Self-Certification Scheme, introduced in 2002, individuals and organisations can self-certify that their work complies with the Building Regulations without the need to submit a Building Notice – avoiding local authority fees. There is a Competent Persons Register, which includes many installers of domestic equipment and facilities.

The contractor is required to give notice of particular critical points within the construction process, triggering a visit from the building control officer who will advise and comment upon matters relating to the building regulations, for example, foundations, drainage, structure, fire protection and so on. It is common for such an inspector to issue directions, but the contractor must, before acting upon them, refer any such direction to the architect for clarification and confirmation as a formal 'instruction'. The architect may wish to consider the inspector's directions for several reasons:

- they may compromise functionality;
- they may have an effect on the programme;
- they may have cost implications;
- they may compromise other agreements, for example, party wall, planning;
- there may be preferable, alternative solutions; and
- there may be the option to discuss with the building control officer whether the directions given are really necessary.

A local authority environmental health officer may wish to inspect the site on the grounds of noise or dust pollution.

An officer from the Health and Safety Inspectorate (HSI) is likely to visit the site to inspect such things as CDM compliance, site access, general working conditions and scaffolding to ensure that they meet relevant safety criteria. The HSI inspector may also ask to see the contractor's Risk Register under CDM.

Subject to the type of building, a Fire and Civil Defence Authority officer may visit to ensure that the means of escape are acceptable, alarms are audible, fire doors adequate and so on.

In addition to such statutory officials, it may be that others seek inspections, for example, NHBC inspectors, surveyors acting for adjoining owners or for insurers and/or the employer's financial backers.

Whoever visits the site, it is important that the contractor verifies the purpose of their visit and checks whether it is advisable for any other party to accompany them. The contractor should always record the visit and make a brief report of the reason for it and the outcome.

Records and reports

Where a clerk of works is employed, he should be required to keep a daily diary of all matters affecting the project. This diary, to be handed to the architect at the end of the project, should in the interim form the basis of weekly reports to the architect. A standard form of report is available from the Institute of Clerks of Works. A completed copy is given as Example 27.2 at the end of this chapter.

Such reports cover:

- the number of men employed daily, in the various trades;
- weather reports and particulars of time lost due to adverse weather conditions;
- length of any work stoppages;
- visitors to site including statutory inspectors;
- delays including action taken to remedy defective work;
- site directions and architect's instructions issued;
- drawings/information received on site;
- drawings/information required on site;
- plant/materials delivered to or removed from site;
- general progress in relation to the programme;

- general report including a summary of work proceeding; and
- comments on site conditions/cleanliness/health and safety.

The clerk of works' diary and weekly reports constitute a most useful record of events, which may be referred to should disputes subsequently arise. In the absence of a clerk of works, the contractor should be required to compile a similar factual record of conditions and activities on site, and the design team should maintain their own records of site observations including plant, labour, activity and progress, to augment or amplify the contractor's records.

A copy of the contractor's programme should be kept in the office of the clerk of works, and the actual progress should be checked and recorded against this programme every week.

It is the job of the clerk of works to keep records of any departures from the production information so that, on completion, architects have all the information necessary to enable them to issue to the employer an accurate set of 'as issued for construction' or even 'as-built' drawings of the finished building. The distinction between these two categories of record drawings is significant as providing 'as built' drawings may require a detailed post-completion survey of the work. These records are particularly important where the work is concealed, as for example with foundations, the depth of which may vary from that shown on the original production information.

Progress photographs of the work also form valuable records if taken regularly. This is probably best arranged in conjunction with the contractor so that the whole building team can have the benefit of them.

If reports are transmitted by email, hard copies should be made for enduring record purposes. Those receiving reports should confirm receipt.

Samples and testing

Architects may call for samples of various components and materials used in the building to be submitted for approval, in order that they can satisfy themselves that they meet the specified requirements of the client and, where applicable, the local authority. Some of the items, of which samples would normally be required, are as follows:

- external cladding materials such as facing bricks, artificial and natural stone, precast concrete, marble, terrazzo, slates and roofing tiles;
- concrete used for flooring where no structural engineer is employed by the client;
- internal finishes such as timber, joinery, ironmongery, floor tiles and wall tiles; and
- services such as plumbing components, sanitary goods and electrical fittings.

In addition to samples of individual components or materials, architects will often require sample panels to be prepared on the site to enable them to judge the effect of the materials in the situations in which they will be used, for example, a panel of facing bricks to demonstrate a particular bond or pattern, the colour of mortar and type of pointing.

It is also quite normal to require laboratory tests of basic materials such as concrete. The testing of concrete should be carried out on a regular basis. Cubes should be cast from each main batch of concrete used and each cube should be carefully labelled and identified. An approved laboratory should carry out the tests, and the test reports should be submitted by the contractor to the architect or structural engineer. Full instructions for such testing procedures are usually included in the specification.

The crushing strength of bricks may also be tested in a laboratory, although unless the design requirements are particularly stringent a certificate from the manufacturer giving their characteristics may well suffice.

If required, manufacturers of external cladding will provide test rigs not only to investigate and demonstrate the appearance and structural performance of their products, but also to test for water penetration under simulated wind pressure.

British Standard Specifications, Agrément Certificates and Codes of Practice are specified for many building materials, components and processes, and in carrying out tests, it is essential to refer to the appropriate standard or code to ensure that the requirements are complied with. In addition, the British Standard Specifications set down acceptable tolerances for manufactured goods. Copies of all relevant standards and codes should always be kept on site by the clerk of works or the contractor.

The architect is empowered under most forms of contract to order tests and also the removal of defective work from site – a notice merely condemning the work is insufficient, and clauses of contracts (e.g. clause 3.17 of the Contract) are formulated to ensure fair play.

Considerate constructors scheme

Constructing a building makes a significant impact on the quality of life in the environment around the site. The impact is particularly significant in city centres, and no more so than in the City of London. In 1987, the Corporation of London pioneered what it called the Considerate Contractors Scheme and, in 1997, the Construction Industry Board launched a similar scheme, known as the Considerate Constructors Scheme. This latter scheme is a voluntary code of practice, which seeks to encourage building, demolition and civil engineering contractors to carry out their operations safely and with consideration to the well-being of others. The Considerate Constructors Scheme, which is operated by the Construction Confederation, has now been adopted across the country. Annual awards are made under the scheme to recognise and reward commitment on the part of contractors to raise standards of site management, safety and environmental awareness beyond their statutory duties. The Code of Practice for the Considerate Constructors Scheme includes the following criteria:

- **Consideration** – All activities are to be carried out with positive consideration for the needs of traders and businesses, site personnel and visitors, and

Part IV

the general public. Special attention is to be given to the needs of those with disabilities.

- **Environment awareness** – Noise from construction work, site personnel and other sources is to be kept to a minimum at all times. Local resources should be used wherever possible. Waste management and the avoidance of pollution by recycling surplus materials are to be encouraged.
- **Cleanliness** – The working site, safety barriers, lights and signs are to be maintained in a clean and safe condition. Surplus materials and rubbish must not be allowed to accumulate on site or spill over onto adjacent property. Dust from all activities is to be kept to a minimum.
- **Good neighbours** – Full and regular consultation with neighbours regarding programming and site activities is to be maintained from pre-start to completion of activities.
- **Respect** – All site personnel are to wear respectable and safe dress, appropriate to the weather conditions, and are to be instructed in dealing with the general public. Lewd or derogatory behaviour and language are not to be tolerated. The contractor is to take pride in the management and appearance of the site and surrounding environment.
- **Safety** – Movements of site personnel and vehicles are to be carried out with great care and consideration for traders and businesses, site personnel and visitors, and the general public. No building activity is to be a security risk to others.
- **Responsibility** – The contractor is to ensure that all persons working on the site understand and implement the Code of Practice.
- **Accountability** – Posters are to be displayed clearly around the site, giving names and telephone numbers of staff who can be contacted to respond to issues raised by those affected.

Site safety

Health and Safety Policy

We live and work in a safety-conscious and highly regulated age. Construction sites should be safe places for those who work on them, for those who are site neighbours, for those who maintain buildings and for the users themselves.

The route to site safety is through the CDM 2015 and a number of Health and Safety regulations. Architects have specific duties in relation to CDM 2015 and Health & Safety.

CDM 2015: If the project is notifiable, a principle designer will have been appointed at or around Stage C. The client will have been informed by the architect of their duties under CDM, which for most clients, the architect usually performs. The Risk Register will have been compiled during the design process. The Architect is required under CDM 2015 to ensure that all workplaces meet the requirements of the Workplace (Health and Safety Regulations) where there is some overlap with Building Regulations. The architect is under a duty to identify and record significant and unusual hazards, some of which may only occur once work has started on site. The Risk Register is handed to the client at Practical Completion. In simple terms, the architect must observe site conditions as the work progresses and report to officials any potential breach of safety regulations.

Generally, issues to be considered by the contractor when drafting his safety policy statement are many and various:

- The Risk Register under CDM 2015 should be a live document,
- The contractor's Health and Safety Policy Statement to which the architect will refer on site should express a commitment to health and safety and make clear the contractor's obligations towards employees.
- The statement should record which person in authority is responsible for ensuring that it is implemented and regularly reviewed.
- The statement should be signed and dated by a partner or senior director.
- The statement should take account of the views of managers, supervisors, safety representatives and the safety committee.
- The duties set out in the statement should be discussed in advance with the people concerned and accepted by them to ensure that they understand how their performance is to be assessed and what resources they have at their disposal.
- The statement should make it clear that co-operation on the part of all employees is vital to the success of the contractor's health and safety policy.
- The statement should record how employees are to be involved in health and safety matters, for example by being consulted, by taking part in inspections and by sitting on a safety committee.
- The statement should show clearly how the duties for health and safety are allocated and describe the responsibilities at different levels.
- The statement should record who is responsible for the following matters (including deputies where appropriate):
  - reporting investigations and recording accidents
  - fire precautions, fire drill, evacuation procedures
  - first aid
  - safety inspections
  - the training programme
  - ensuring that legal requirements are met, for example regular testing of lifts
  - notifying accidents to the health and safety inspector

Site safety

- Delivery access, timing and loading restrictions
- Storage compound location
- Inflammable product storage
- De-commissioning, storage or disposal of any existing equipment
- Position of tool store, welfare facilities and shared facilities
- Security and control of the access to high-risk work areas, machinery rooms, control equipment and high voltage switch gear
- Scaffolding and protection arrangements
- Permits-to-work for flame producing or oxy-acetylene cutting gear

Part IV

- Control of Substances Hazardous to Health (COSHH) assessments of dusty works or substances used in occupied or shared work areas
- Temporary lighting requirements or specification of power locations and usage
- Risk assessments of work operations, relevant in detail and scale to the project in progress
- Method statements relating to the implementation of high risk elements of the work, giving details and descriptions relevant to the size and scale of the works
- Commissioning logs and partial hand-over arrangements
- Lift usage restrictions/authorised user list
- Arrangements for dealing with fire prevention
- Training and competence certificates
- First aid arrangements
- Emergency contact list
- Named management and their respective responsibilities
- Details of how the archived information will relate to the safety file and what operation and maintenance information will be given to the principal contractor during the project
- Keeping the workplace, including staircases, floors, ways in and out, washrooms and so on in a safe and clean condition by cleaning, maintenance and repair.

Plant and substances

- Maintenance of equipment such as tools and ladders in a safe condition
- Maintenance and proper use of safety equipment such as helmets, boots, goggles, respirators and so on
- Maintenance and proper use of plant, machinery and guards
- Regular testing, maintenance and emergency repair of lifts, hoists, cranes, pressure systems, boilers and other dangerous machinery
- Maintenance of electrical installations and equipment
- Safe storage, handling and, where applicable, packaging, labelling and transport of dangerous substances
- Controls on work involving harmful substances such as lead and asbestos
- The introduction of new plant, equipment or substances into the workplace – by examination, testing and consultation with the workforce.

Other hazards

- The wearing of ear protection, and control of noise at source
- Preventing unnecessary or unauthorised entry into hazardous areas
- The lifting of heavy or awkward loads
- Protecting the safety of site staff against assault when handling or transporting the employer's money or valuables
- Special hazards to site staff when working on unfamiliar sites, including discussion with site manager where necessary
- Control of works transport, such as forklift trucks, by restricting use to properly trained, experienced operatives.

Emergency procedures

- Ensuring that fire exits are marked, unlocked and free from obstruction
- Maintenance and testing of fire-fighting equipment, fire drills and evacuation
- Names and location(s) of persons responsible for first aid and any deputies, and the location of all first aid boxes.

Communication procedures

- Giving site staff information about their general duties under HASAW and specific legal requirements relating to their work
- Giving site staff necessary information about all substances, plant, machinery and equipment with which they come into contact
- Discussing with sub-contractors, before they come on site, how they can plan to do their job with a view to minimising risk, and how to deal with unavoidable hazards that they may create for site staff and that their own employees may encounter on site.

Training procedures

- Ensuring that all employees, supervisors and managers have the necessary training to enable them to work safely and to carry out their health and safety responsibilities efficiently.

Supervision procedures

- Supervising site staff so far as is necessary for their safety – especially young workers, new employees and employees carrying out unfamiliar tasks.

Checking procedures

- Making regular inspections and checks of the workplace, machinery, appliances and working methods.

Fire precautions on site

The architect on site should be aware of fire and life safety provisions and take all reasonable and necessary steps to mitigate loss and injury. Fire is a serious and ever-present danger on site and is a key issue when considering site safety. Fires spread rapidly and can endanger life and cause massive damage to property, including collateral damage to neighbouring buildings and equipment. They are a more frequent occurrence than might be thought, there being several construction site fires of some magnitude somewhere every day. Fire precautions depend on the nature and location of the project. Erecting a simple steel-framed building in a rural location will require only simple precautions because fire risks, of occurrence and effect, are minimal in such circumstances. Higher risk work, for example

Part IV

refurbishing floors in an occupied office block, will need many more precautions because the risk of fire occurring, the difficulties of escape and the potential for injury and loss of life are all that much greater (the sites of conservation projects are particularly hazardous). In any event, fire alarm points and local extinguishers (to enable escape, not to fight the fire) must be available throughout the site to comply with regulations.

Regulatory control

The CDM Regulations 2015 require contractors to take measures both to prevent fires occurring and to ensure that everyone on site, including visitors, is protected if fires happen. From the outset, CDM2015 requires all those designing, planning and constructing projects to take account of construction site fire safety in their thinking at all stages.

For larger projects, the Joint Code of Practice on the Protection from Fire of Construction Sites and Buildings Undergoing Renovation (Joint Fire Code) is published by the Construction Confederation and the Fire Protection Association with more detailed and prescriptive requirements.

The Joint Fire Code

This Code applies to works of £1M in value and over. Works of £25M in value and over are termed 'Large Projects' and for these the Code provides additional requirements.

In the sixth edition of the Joint Fire Code, which was published in 2006, it states in clause 2 that:

'Non-compliance with the Code by the Construction Industry, by those who procure construction and by construction industry professionals could result in insurance ceasing to be available or being withdrawn resulting in a possible breach of construction contracts which require the provision of such insurance'.

Failure to comply with the Joint Fire Code might lead to the insurance underwriters withdrawing cover for a project. This would be a great risk to contractor and client alike. If a project were not covered by insurance, and the contractor were to inadvertently start a fire, the client might have to begin legal proceedings to recover damages. As most contractors operate using a fluid cash flow, there would be no guarantee that the contractor could sustain an action for damages where reinstatement costs are high. It is likely that the contractor would become insolvent and cease trading, leaving the client to carry the loss.

Clauses 6.14 to 6.16 of the SBC/XQ provide for compliance with the Joint Fire Code by both the client and the contractor when it applies to their particular contract. The SBC also makes provision to ensure that any 'Remedial Measures' required by an insurer to rectify a breach of the Joint Fire Code are carried out timeously.

Part IV

An example of the code's requirements is the use on site of high-quality flame retardant protection materials approved to Loss Prevention Standard LPS 1207.

Preventing fire

Basic precautions to prevent fires on construction sites are as follows:

- Store LPG cylinders and other flammable materials properly. LPG cylinders should be stored outside buildings in well-ventilated and secure areas. Other flammable materials such as solvents and adhesives should be stored in lockable steel containers. Care must be taken to avoid leaks.
- Carefully control hot work such as welding by using formal permit-to-work systems.
- Do not leave tar boilers unattended.
- Keep an orderly site and make sure rubbish is cleared away regularly.
- Avoid unnecessary stockpiling of combustible materials such as polystyrene.
- Take special precautions in areas where flammable atmospheres may develop, for example when using volatile solvents or adhesives in enclosed areas.
- Avoid burning waste materials on site wherever possible. Never use petrol or similar accelerants to start or encourage fires.
- Make sure everyone obeys site rules on smoking.

Means of escape

Construction sites can pose particular problems because the routes in and out may be incomplete or subject to change as the project advances and obstructions may be present. Open sites usually offer plentiful means of escape and special arrangements are unlikely to be necessary. In enclosed buildings, people can easily become trapped, especially where they are working above or below ground level. In such cases, means of escape need careful consideration. The contractor must make sure that:

- Wherever possible, there are at least two escape routes in different directions.
- Travel distances to safety are reduced to a minimum.
- Enclosed escape routes, for example corridors or stairwells, can resist fire and smoke ingress from the surrounding site. (Where fire doors are needed for this, the contractor must ensure that they are provided and kept closed; self-closing devices should be fitted to doors on enclosed escape routes.)
- Escape routes and emergency exits are clearly signed.
- Escape routes and exits are kept clear.
- Emergency exits are never locked when people are on the site.
- Emergency lighting is installed if necessary to facilitate escape. (This is especially important in enclosed stairways in multi-storey structures, which will be in total darkness if the normal lighting fails during a fire.)
- An assembly point is identified where everyone can gather and be accounted for.

Part IV

Fire-fighting equipment

The equipment needed depends on the risk of fire occurring and the likely consequences if it does. It can range from a single extinguisher on small low-risk sites to complex fixed installations on large and high-risk sites. In any event the contractor must ensure that:

- Fire-fighting equipment is located where it is really needed and is easily accessible.
- The location of fire-fighting equipment and instructions on its use are clearly indicated.
- The right sorts of extinguishers are provided for the type of fire that could occur. (A combination of water or foam extinguishers for paper and wood fires and CO_2 extinguishers for fires involving electrical equipment is usually appropriate.)
- The equipment provided is maintained and is operational. (A competent person, normally from the manufacturer, should check fire-fighting equipment regularly.)
- Those carrying out hot work have appropriate fire extinguishers with them and know how to use them.

Emergency plans

The purpose of emergency plans is to ensure that everyone on site reaches safety if there is a fire. Small and low-risk sites only require very simple plans, but higher risk sites will need more careful and detailed consideration. An emergency plan should:

- be available before work starts on site;
- be up to date and appropriate for the circumstances concerned;
- make clear who does what during a fire;
- be incorporated into the construction phase health and safety plan under CDM2015; and
- work if it is ever needed.

Providing information

Fire action notices should be clearly displayed in locations where every site operative can see them. For example, most notices are displayed at site entrances, plant and material distribution points, canteen areas and wash room facilities. Site operatives should be made aware of the locations where information can be found. This is usually done during a site induction meeting prior to an individual commencing work on site.

Record keeping

Keeping accurate and detailed records of matters related to the works on site is very important, for the control of the works, for Health and Safety reasons, for use in any delay analysis and in the resolution of any future dispute. The architect should make sure that the following documentation is created and retained on site by the contractor during the course of his works:

- Register of those attending site including time of arrival on and departure from site.
- Diary of site events including weather.
- Incident book to record any injuries or illness.
- Information including welfare arrangements, key contacts, emergency procedures, for example, fire evacuation, spillages, first aid, emergency services shut-off points.
- Training record for site staff.
- Asbestos register (if relevant).
- Test certificates for any equipment running on site, for example, ventilation.
- Permits to work (if relevant).

The architect will produce documentation to record events, and the principal documents are as follows:

- Minutes of site meetings including those attending, points raised, action points and records of actions completed and outstanding, applications for delay and requests for instructions. The minutes must record the names of those responsible for completing actions and must be written and distributed without delay.
- Architect's instructions issued under the construction contract, which may amplify or vary the contracted scope of work and which may have an implication on cost and time. Instructions may also confirm those given orally.
- Variations.
- Removal of non-compliant work.
- Notice of non-compliance with an instruction.
- Interim Valuations and Interim Certificates.
- Schedules of Defective Work and Certificates of Making Good Defects.
- Certificate of Partial Possession.
- Certificate of Practical Completion.
- Final certificate.
- Extensions of time.

It is not unknown for the contractor to receive a certificate of Practical Completion and immediately send to the architect a major claim for extensions of time. It is important to record events on site meticulously, so that in significant cases, experts can analyse events forensically with a high degree of accuracy. Good minutes of site meetings are a key tool, enabling the architect to manage jobs efficiently and effectively.

Part IV

This list should be amplified or adapted according to the nature of the project and may serve the architect or engineer. Some of the items should be observed jointly with other consultants.
General:

- In all cases check that the work complies with the latest drawings and specification, with the latest requirements of the statutory undertakers and with the building regulations. Ensure that all information is complete.

Preliminary works:

- scaffolding and the Building Act 1984
- location of workmen's canteens, builder's offices and so on
- suitability and location of clerk of works' or site architect's office
- removal of top soil and location of spoil heaps
- perimeter fencing or hoardings
- protection of rights of way
- protection of trees and other special site features
- protection of materials on site
- party-wall agreements and protection of adjoining property
- site security generally
- agreement on bench mark or level pegs
- agreement on setting out (responsibility of the contractor)

Demolition:

- extent
- adequacy of shoring
- preservation of certain materials and special items (to be listed in the specification or bills of quantities)

Excavation and foundations:

- widths of trenches
- depths of excavations
- nature of ground in relation to trial hole report
- stability of excavations
- pumping arrangements
- risk to adjoining property and general public
- quality of concrete and thickness of beds
- suitability of hardcore (freedom from rubbish)
- suitability of sand and ballast (freedom from loam and correct grading)
- damp-proof membranes and tanking
- ducts, drains or services under building
- size, bending, spacing and placing of reinforcement
- suitability of material for and consolidation of backfilling
- depths of piles and driving/boring conditions

Drainage:

- depths of inverts and gradients of falls
- thickness and type of bed and jointing of pipes
- quality of bricks for manholes and rendering thereto
- testing of drains and manholes
- need for a build-over agreement from the local water authority where work covers sewer piping.

Example 27.1   Standard checklist for site inspections.

Brickwork, blockwork and concrete masonry:
- approval of sample panels of facings and fairface work
- quality and colour of mortar and pointing
- test report on crushing strength where necessary
- BS certificates on load-bearing blocks
- position and type of wall ties
- cleanliness of cavities and wall ties in external walls
- setting-out and maintenance of regular vertical and horizontal joints
- type, quality and placing of damp proof courses
- setting-out and fixing of door frames, windows and so on
- bedding and levels of lintels over openings
- expansion joints

In-situ concrete:
- setting out and stability of shuttering
- shuttering type to achieve finish specified
- setting out of reinforcement, fixings, holes and water bars
- mix for and procedure of taking test cubes
- curing of concrete and striking of shuttering
- moisture barriers and foundation layers
- vibration
- use of additives

Precast concrete:
- size and shape of units
- quality of fit and finish
- position of fixings, holes and so on
- damage in transit and erection

Carpentry and joinery:
- freedom from loose knots, shakes, sapwood, insect attack and so on
- dimensions within permissible tolerances
- application of timber preservatives and primers
- storage, stacking and protection from weather
- jointing, bolting, spiking and notching of carpenter's timber
- spacing of floor and ceiling joists and position of trimmers
- spacing of battens, position of noggings
- weather throatings and cills to doors, windows and so on
- jointing, machining and finish of manufactured joinery
- fire rating

Roofing:
- pitch of roof
- spacing of rafters and tile battens
- approval of under felt
- approval of roofing materials and fixings
- pointing to verges and bedding of ridges and so on
- falls to outlets on flat roofs

Example 27.1 (*Cont.*)

Part IV

- thickness and fixing of insulation under finish
- eaves details and ventilation
- correct formation of flashings and so on
- ventilation

Cladding:
- vapour barriers and insulation
- regularity of grounds for sheet materials
- location and quality of fixings
- laps, tolerances and positions of joints
- setting-out and jointing of mullions and rails
- handling and protection of panel materials
- specification and application of mastic
- flashings, edge trims and weather drips
- entry and egress of moisture
- prevention of electrolytic action
- location and type of movement joints

Steelwork:
- sizes of steel
- tightness (torque) of attachment bolts
- rivets and welding
- position of members
- plumbing, squaring and levelling of steel frame
- priming and protection

Metalwork:
- sizing and spacing of members
- galvanising and rustproofing
- stability of supports, including caulking or plugging
- isolation from corrosive materials and so on

Plumbing and sanitary goods:
- ensuring that sanitary goods are free from cracks and deformities
- location, venting and fixing of stack pipes
- falls to waste branches
- use of traps
- jointing of pipes
- smoke and/or water tests
- location and accessibility of valves, stop-cocks
- drain-down cocks at lowest points
- access to traps and rodding eyes

Heating, hot water and ventilation installations:
- types of boiler, cylinder, tanks, fans and so on
- types of pipes
- position and type of stop valves
- position of pipe runs and ventilation ducting
- insulation of pipework and ducting
- identification and labelling of pipes, valves and so on

Example 27.1   *(Cont.)*

Electrical installation:
- types of switchgear, distribution board, motor drives and so on
- types of switches, socket outlets, fuses, cables and so on
- location of outlet points
- earthing of installation
- lightning conductor installation
- runs of cables and quality of connections and so on
- labeling and identification of switchgear and so on

Specialist installations:
- drawings of specialist installations
- drawings of builder's work in connection with specialist installations
- power and plant to be provided
- access for equipment and provision of adequate working space
- temporary support, loading on structure, lifting tackle and so on
- attendance on site and sequence of work

Paving and floor tiling:
- approve materials
- quality of screeds to receive flooring
- junctions of differing floor finishes
- regularity of finish
- falls to gulleys and so on
- skirting and coves
- expansion joints
- types of tile bedding and grouting materials

Plastering and wall tiling:
- storage of materials
- plaster mixes
- preparation of surface
- true surfaces and arrises
- fixing of plasterboard
- filling and scrimming of joints in plasterboard
- adequate hacking of or bonding plaster on concrete
- regularity of finish
- types of tile bedding and grouting materials.

Suspended ceilings:
- type of suspension and tile
- height of ceiling and setting out
- location and co-ordination of light fittings, sprinkler outlets, ventilation grilles and so on
- access panels
- finish trim for curtains, blinds and so on
- fire barriers

Glazing:
- quality of glass and freedom from defects
- integrity of sealed units
- structural capability

Part IV

Example 27.1   (Cont.)

- type and/or thickness
- depth of rebates
- glazing compound, fixing of glazing beads and so on

Painting and decorating:
- approval of materials
- preparation of surfaces and freedom from damp
- ensuring that partly concealed surfaces are properly finished
- ensuring that finished work is free from runs, brush marks and so on
- opacity of finish and so on

Cleaning down and handing over:
- windows cleaned and floors scrubbed
- sanitary goods washed and flushed
- painted surfaces immaculate
- doors correctly fitted, windows not binding or rattling
- ironmongery complete and locks and latches operating correctly
- correct number of and suiting of keys and security cards
- connection of services, provision of meters and so on
- commissioning of all mechanical engineering plant, balancing of air-conditioning and so on
- plant maintenance manuals, plant room service diagrams and so on
- operation of security, communication and fire protection systems
- removal of protective tapes and films and so on
- cleaning or replacement of air filters in the HVAC system
- cleaning of lighting reflectors where uncovered

Example 27.1 *(Cont.)*

Chapter 28
Instructions

Provided that the procedures set out in the preceding chapters are followed, it should be possible to have available complete sets of drawings, specification notes and quotations from sub-contractors and suppliers when the tender documents are prepared. This will, in turn, mean that, as soon as the contract is placed, contractors can be handed all the necessary information to enable them to build the project.

Architect/contract administrator's instructions

However, the ideal circumstances outlined previously are not the norm, and even when fully finalised information is available for incorporation into the contract documents, it will still be necessary, from time to time, for the architect/contract administrator to issue further drawings, details and instructions. These are collectively known as architect's instructions. The standard building contract (SBC) conditions of contract set out those matters in connection with which the architect is empowered to issue such instructions to the contractor, and these are as follows:

Clause	Description
2.10	Levels and setting out of the Works
2.12	Further drawings, details and instructions
2.15	Notice of discrepancies, etc.
2.16	Discrepancies in CDP documents
2.17	Divergences from Statutory Requirements
2.38	Schedules of defects and instructions
3.12	Instructions other than in writing
3.14	Instructions requiring Variations

The Aqua Group Guide to Procurement, Tendering and Contract Administration, Second Edition.
Edited by Mark Hackett and Gary Statham.
© 2016 John Wiley & Sons, Ltd. Published 2016 by John Wiley & Sons, Ltd.

3.15   Postponement of work
3.16   Instructions on Provisional Sums
3.17   Inspection – tests
3.18   Work not in accordance with the Contract
3.19   Workmanship not in accordance with the Contract
3.21   Exclusion of persons from the Works
3.22   Instructions on antiquities
5.3   Variation Quotation
6.5   Contractor's insurance of liability of Employer
6.16   Breach of Joint Fire Code – Remedial Measures
Schedule 2   Paragraph 4 – Acceptance of quotation
             Paragraph 5.1 – Non-acceptance of the quotation
Schedule 8   Paragraph 3.3 – Cost savings and value improvements.

Clause 3.10 of the SBC provides the contractor with the mandatory obligation to comply with an architect/contract administrator's instruction and circumstances in which the contractor needs not to comply with those instructions. Clause 3.12, however, deals with any instructions other than in writing. These can be summarised as follows:

- any instruction issued by the architect/contract administrator must be in writing;
- should the architect/contract administrator issue an instruction other than in writing, it is of no effect unless, within 7 days of such an instruction being given:
  - it is confirmed in writing by the architect/contract administrator, at which time it becomes effective, or
  - it is confirmed in writing by the contractor and is not dissented from by the architect/contract administrator within a further period of 7 days, at which time it becomes effective; and
- if neither the architect/contract administrator nor the contractor confirms an instruction issued other than in writing but the contractor complies with the instruction, the architect/contract administrator can, at any time up to the issue of the final certificate, confirm that instruction in writing with retrospective effect.

It is worthy of note that SBC Clause 3.13 allows the contractor to question the contractual validity of any architect/contract administrator's instruction and requires the architect/contract administrator to answer 'forthwith'. It should also be noted that, if the contractor does not comply with a valid architect/contract administrator's instruction, Clause 3.11 allows the employer to employ others to give effect to the instruction and recover all costs so incurred from the defaulting contractor.

The SBC also makes provision for the architect/contract administrator to issue instructions to the quantity surveyor, and these are as follows:

Clause	Description
4.18.2	To prepare a statement specifying the details of the Retention deducted in arriving at the amount stated as due in interim certificates.
4.19.1	To prepare a statement specifying the amount of Retention that would have been deducted had it been permissible to deduct Retention (where it is stated in the Contract Particulars that a bond is required rather than deducting Retention).
4.23	To ascertain the amount of loss and/or expense incurred by the contractor.

Clerk of works' directions

If a clerk of works is employed on the project and issues any directions to the contractor, such directions are effective only if they are issued in regard to a matter in respect of which the architect/contract administrator is empowered to issue instructions, and also if they are confirmed in writing by the architect/contract administrator within 2 working days of the direction being given (not, it is noted, within 7 days as is provided for the confirmation by architects/contract administrators of their own instructions issued other than in writing).

Format and distribution of instructions

It is good practice for all instructions from the architect/contract administrator to the contractor to be issued or confirmed on standard forms. An example of such a form, which is published by RIBA publishing, is given as Example 28.1 at the end of this chapter.

It is essential that instructions should be clear and precise, and where revised drawings are issued, the date and reference of the particular revision should be specifically referred to. Instructions emanating from other members of the design team must not be given directly to the contractor but must be issued via the architect/contract administrator as architect/contract administrator's instructions. Copies of architect's instructions should be distributed to all of the following:

- contractor;
- employer or employer's representative;
- project manager;
- principal designer;
- quantity surveyor;
- other members of the design team (e.g. structural engineer, services engineer);
- clerk of works; and
- any sub-contractor affected by the instruction.

Part IV

				Architect's Instruction

Issued by: Ivor Barch Associates
address: Prospects Drive, Fairbridge

Employer: Cosmeston Preparatory School
address: Fairbridge

Contractor: L&M Construction Ltd
address: Ferry Road, Fairbridge

Works: New School Library
situated at: Park Street, Fairbridge

Job reference: IBA/94/2

Instruction no: 4

Issue date: 18 February 2014

Sheet: 1   of 1

Contract dated: 14 December 2013

Under the terms of the above-mentioned Contract, I/we issue the following instructions:

		Office use: Approximate costs	
		£ omit	£ add
1.	INCOMING GAS MAIN Accept the quotation ref. no 8438/63 dated 6 January 2014 received from Eurogas in the sum of £248.00 for the new incoming gas main and supply and installation of gas meter.	400.00	248.00
2.	HIP TILES OMIT: Farland concrete third round hip tiles, bill of quantities ref. 5/15E. ADD in leiu: Red Bank 300mm long Terracotta third round segmental ridge tiles. list no. 259.	372.00	514.00
3.	PICTURE HANGING FACILITIES IN ACTIVITIES ROOM Supply and fix aluminium picture rail approx. 12m long and 100no. type (b) U-shaped hooks obtainable from Library Aids of Thawbridge. Picture rail to be fixed in location shown on attached drawing no. IBA/94/20/61 at 2050mm from finished floor level.	-	80.00

To be signed by or for the issuer named above

Signed _RwP.pe_____

772.00	842.00

Amount of Contract Sum	£ 506,096.00
± Approximate value of previous Instructions	£ 500.00
Sub-total	£ 505,596.00
± Approximate value of this Instruction	£ 70.00
Approximate adjusted total	£ 506,666.00

Distribution					
	☐ Contractor	☐ Structural Engineer	☐ CDM Co-ordinator	☐	
	☐ Employer	☐ M&E Consultant	☐	☐	
	☐ Quantity Surveyor	☐ Clerk of Works	☐	☐ File	

for SBC / IC / ICD / MW / MWD   CONTRACT ADMINISTRATION FORMS   © RIBA Publishing 2011

Example 28.1 Architect's instruction. Reproduced by permission of Royal Institute of British Architects.

Chapter 29
Variations and Post-Contract Cost Control

Variations

It is important not to confuse architect/contract administrator's instructions with variations. Not all architect/contract administrator's instructions give rise to a variation. In the standard building contract (SBC), clause 5 provides the following definition of a variation:

5.1   The term 'Variation' means:

    5.1.1   the alteration or modification of the design, quality or quantity of the Works, including:

        5.1.1.1   the addition, omission or substitution of any work;

        5.1.1.2   the alteration of the kind or standard of any of the materials or goods to be used in the Works;

        the removal from the site of any work executed or Site

        5.1.1.3   Materials other than work, materials or goods which are not in accordance with the Contract;

    5.1.2   the imposition by the Employer of any obligations or restrictions in regard to the matters set out in this clause 5.1.2 or the addition to or alteration or omission of any such obligations or restrictions so imposed or imposed by the Employer in the Contract Bills or in the Employer's Requirements in regard to:

        5.1.2.1   access to the site or use of any specific parts of the site;

        5.1.2.2   limitations of working space;

        5.1.2.3   limitations of working hours; or

        5.1.2.4   the execution or completion of the work in any specific order.

The Aqua Group Guide to Procurement, Tendering and Contract Administration, Second Edition.
Edited by Mark Hackett and Gary Statham.
© 2016 John Wiley & Sons, Ltd. Published 2016 by John Wiley & Sons, Ltd.

The matters set out in SBC clause 5.1.2 relate to the working conditions imposed by the employer, particulars of which will have been given in the bills of quantities in accordance with Table 1 of NRM2 (previously Section A of SMM7R) or in the documents containing the Employer's Requirements. It should be noted that whilst there is a general obligation, under SBC clause 3.10, for the contractor to comply forthwith with all architect/contract administrator's instructions, SBC clause 3.10.1 allows that the contractor need not comply with an architect's instruction requiring a variation within the meaning of SBC clause 5.1.2 insofar as he has given the architect/contract administrator reasonable, written objection to such compliance. The reason for giving the contractor this important right of objection may well be in recognition of the possibility that such a variation could so fundamentally affect the basis on which the contractor tendered for the project that the valuation of variations and ascertainment of direct loss and/or expense provisions within the SBC would not provide adequate recompense.

Valuing variations

All variations, other than those for which a contractor's Schedule 2 Quotation has been accepted pursuant to SBC clause 5.3, fall to be valued by the quantity surveyor in accordance with the Valuation Rules at SBC clauses 5.6–5.9. These clauses prescribe the five basic methods for the valuation of all work executed by the contractor as the result of architect/contract administrator's instructions as to the expenditure of provisional sums and the execution of work for which an approximate quantity is included in the contract bills and variations.

The method of valuation to be adopted by the quantity surveyor depends on the nature of the work and the conditions under which it has to be executed. Each basic method of valuation and the nature of the work to which it is to be applied are summarised in Table 29.1.

The method of valuing variations relating to the contractor's designed portion (CDP) is made broadly similar insofar as it relates to work set out in the CDP analysis rather than that contained within the contract bills (Table 29.2).

It must be emphasised that, under SBC clause 5.10.2, no allowance is to be made in any valuation carried out under SBC clause 5.2 for any effect upon the regular progress of the works or for any other direct loss and/or expense for which the contractor would be reimbursed by payment under any other provision of SBC. A contractor should seek to recover loss and/or expense under SBC clause 4.23, which deals with matters arising from relevant events that materially affect the regular progress of the works.

Nevertheless, pursuant to SBC clause 5.3, the architect/contract administrator may request the contractor to provide a quotation for a variation in accordance with SBC Schedule 2. Subject to the contractor receiving sufficient information in order to provide a quotation, the contractor is required to provide a quotation. That is unless within 7 days of receiving the instruction (or such other period in the instruction or agreed between them), the contractor notifies the architect/contract administrator that he disagrees with the application of the procedure to the particular instruction.

Table 29.1 The five basic methods for the valuation of variations (excluding variations relating to a CDP and Schedule 2 Quotation).

Description of Work	Method of Valuation
1. Additional or substituted work that can properly be valued by measurement, and where the work is of similar character to, is executed under similar conditions as, and does not significantly change the quantity of work set out in the contract bills. Work for which an approximate quantity is included in the contract bills and that quantity is reasonably accurate. Omission of work set out in the contract bills.	Measure in accordance with NRM2. Value at the rates and prices for the work set out in the contract bills. Allow for any percentage or lump sum adjustments in the contract bills. Allow for any addition to or reduction of preliminary items of the type referred to in NRM2 Table 1 providing that no allowance is made for work resulting from an instruction as to the expenditure of a provisional sum in the contract bills for defined work.
2. Additional or substituted work that can properly be valued by measurement and where the work is of similar character to work set out in the contract bills but is not executed under similar conditions thereto and/or significantly changes the quantity thereof. Work for which an approximate quantity is included in the contract bills and that quantity is not reasonably accurate. Any work executed under changed conditions as a result of a variation, compliance with an instruction as to the expenditure of a provisional sum or the execution of work for which an approximate quantity is included in the contract bills.	As 1 above plus a fair allowance for the differences arising from the work not being executed under similar conditions and/or because the variation changes the quantity of the work of similar character in the contract bills.
3. Additional or substituted work that can properly be valued by measurement and where the work is not of similar character to work set out in the contract bills.	Measure in accordance with NRM2. Value at fair rates and prices. Allow for any percentage or lump sum adjustments in the contract bills. Allow for any addition to or reduction of preliminary items of the type referred to in NRM2 Table 1 providing that no allowance is made for work resulting from an instruction as to the expenditure of a provisional sum in the contract bills for defined work.

(continued overleaf)

Part IV

Table 29.1 (*continued*)

Description of Work	Method of Valuation
4. Additional or substituted work that cannot properly be valued by measurement.	Provided that vouchers specifying the time spent on the work, the workmen's names, the plant and the materials employed are delivered to the architect for verification not later than the end of the week following that in which the work is executed, value at the prime cost of such work calculated in accordance with the 'Definition of Prime Cost of Daywork carried out under a Building Contract' issued by the Royal Institution of Chartered Surveyors and the Construction Confederation or the Electrical Contractors' Association or the Heating and Ventilating Contractors' Association, as appropriate, and current at the base date together with percentage additions to each section of the prime cost at the rates set out in the contract bills.
5. Work other than additional, substituted or omitted work. Work or liabilities directly associated with a variation that cannot reasonably be valued by any other method.	A fair valuation.

Table 29.2 The valuation of variations relating to a CDP.

Description of Work	Method of Valuation
1. Additional or substituted work consistent with work of similar character to that set out in the CDP analysis and there is no change in conditions under which the work is carried out and/or a significant change in quantity of the work so set out. Omission of work set out in the CDP analysis.	Valuation consistent with the values set out in the CDP analysis for such work. Allowance for the addition or omission of relevant design work. Allow for any percentage or lump sum adjustments in the contract bills. Allow for any addition to or reduction of preliminary items of the type referred to in NRM2 Table 1, providing that no allowance is made for work resulting from an instruction as to the expenditure of a provisional sum in the contract bills for defined work.
2. Additional or substituted work consistent with work of similar character to that set out to that in the CDP analysis but there is a change in conditions under which the work is carried out and/or a significant change in quantity of the work so set out.	As 1 above plus a fair allowance for the differences arising from the work not being executed under similar conditions and/or because there is a significant change in the quantity of the work so set out.
3. Where there is no work of a similar character set out in the CDP analysis.	A fair valuation. Allowance for the addition or omission of relevant design work. Allow for any percentage or lump sum adjustments in the contract bills. Allow for any addition to or reduction of preliminary items of the type referred to in NRM2 Table 1, providing that no allowance is made for work resulting from an instruction as to the expenditure of a provisional sum in the contract bills for defined work.
4. Additional or substituted work that cannot properly be valued by measurement (insofar as relevant).	Provided that vouchers specifying the time spent on the work, the workmen's names, the plant and the materials employed are delivered to the architect for verification not later than the end of the week following that in which the work is executed, value at the prime cost of such work calculated in accordance with the 'Definition of Prime Cost of Daywork carried out under a Building Contract' issued by the Royal Institution of Chartered Surveyors and the Construction Confederation or the Electrical Contractors' Association or the Heating and Ventilating Contractors' Association, as appropriate, and current at the base date together with percentage additions to each section of the prime cost at the rates set out in the contract bills.

Part IV

Insofar as the contractor notifies the architect/contract administrator of his disagreement within 7 days, he is not obliged to provide a quotation and the variation shall not be carried out. In order for the variation to be carried out, the architect/contract administrator is required to issue another instruction that the variation is to be carried out and it is to be valued as by a valuation (in accordance with the valuation rules).

The Schedule 2 Quotation in respect of a variation allows the contractor to submit a quotation based on valuing all of the effects of a variation. Such quotation may include a claim for loss and/or expense, although such claim should be separately identifiable from other aspects of the quotation. At this juncture, it is important to realise that this is not an opportunity for the contractor simply to value resources such that it can make up for any losses of his own making. The architect/contract administrator and, in particular, the quantity survey must be diligent and investigate all aspects of the contractor's quotation.

The final item to consider concerns SBC clause 5.9 and any change of conditions for other work arising out of: (i) compliance with an instruction for a variation; (ii) compliance with an instruction as to the expenditure of a provisional sum in respect of undefined work; (iii) compliance with an instruction as to the expenditure of a provisional sum in respect of defined work, to the extent that work differs from the description given for such work in the contract bills; (iv) the execution of work covered by an approximate quantity in the contract bills and the extent that there is more or less ascribed to that work in the contract bills. Insofar as there is substantial change in the conditions under which any other work is executed arising out of any of the four previously mentioned circumstances, then such other work shall be treated as if it is subject of an instruction requiring a variation. Any such work treated as subject of a variation is then valued by reference to the valuation provisions.

A point worthy of note is that when the contract rules for valuing variations are brought into play, save in respect of a Schedule 2 Quotation, the quantity surveyor has a unilateral responsibility – the contractor is not involved with the single exception that he is entitled to be present when any measurements are taken. Should the contractor not be satisfied with the quantity surveyor's valuation, the only formal recourse is to use the dispute resolution procedures set out in the contract. Practically, of course, the quantity surveyor usually works closely with the contractor's surveyor so that (more often than not) disputes are avoided and an agreed final account is produced.

Dayworks

Works valued at prime cost on a daywork basis effectively involve a cost-plus method of reimbursement, and from a contractor's viewpoint, this can be viewed an attractive means of securing payment for variations. For work to be valued at prime cost on a daywork basis, the record sheets must be submitted to the architect/contract administrator or his authorised representative, usually the clerk of works, no later than 1 week following the week in which the work was carried out. In addition to being serially numbered, it is essential that the following information be recorded on each sheet:

- the reference of the architect/contract administrator's instruction authorising the work;
- the date(s) when the work was carried out;
- the daily time spent on the work, set against each operative's name;
- the plant used; and
- the materials used.

The architect/contract administrator or his authorised representative should check the accuracy of the records as soon as possible after receiving the sheets and, if the records are found to be correct, countersign them and pass them to the quantity surveyor. The quantity surveyor is then required to follow the valuation rules in order to determine whether variation is to be valued at prime cost. It is worth noting that the architect/contract administrator's signature on a daywork sheet does not constitute a variation. Equally, nor does it commit the quantity surveyor to having to value the work at prime cost. It is for the quantity surveyor to decide which method of valuation is the most appropriate in the circumstances.

It is helpful if the contractor gives the architect/contract administrator advance warning of his intention of recording anything as daywork so that the architect/contract administrator, or the clerk of works on his behalf, can arrange for particular notice to be taken of the resources used.

The RICS provides advice for its members in respect of valuing variations under its guidance note 'Valuing Change' under the headings of General Principles, Practical Applications and Practical Considerations. The guidance note covers contracts such as JCT, New Engineering Contract/Engineering and Construction Contract and FIDIC.

Cost control

Cost control may be defined as the controlling measures necessary to ensure that the project is delivered within the employer's reasonable budget having regard to the risks accepted at the commencement of the project. It is a continuous process and follows on from the pre-contract cost control activities discussed in Chapter 18.

It is essential for the design team to establish at the outset the parameters within which the employer requires the construction costs to be controlled. Usually, employers have a limit on their project expenditure and will insist that this is not exceeded. Nevertheless, occasionally, an employer will have a need to achieve a high-quality end product or to achieve in the quickest time possible, and these criteria are more important to the employer than project cost.

Initially, the authorised expenditure limit for a project will more often than not be the contract sum. However, it is not uncommon for that limit to be varied during construction. For example, a speculative developer may find that a prospective tenant is prepared to pay more for an enhanced specification. In which case, the employer will require the design team to assess the cost and time implications of incorporating the enhanced specification into the project, and if it is economically viable, the employer will approve an increase in the previously authorised expenditure limit.

As most construction projects are complex, it is normal for a contingency amount to be included in the contract sum to cater for expenditure on unforeseen items of work that become necessary during the construction process. The amount of the contingency sum will vary according to the nature of the project and the perceived risk in respect of foreseen and unforeseen items. A refurbishment project would more likely require a larger contingency provision than would a similar value new building project on a green-field site. It would be normal for the contingency sum to equal between 3% and 5% of the contract sum, although it could be more on a particularly risky project. The primary purpose of the contingency sum is to fund additional work that could not reasonably have been foreseen at design stage, for example, additional work below ground; it is not there to fund design alterations, except with the prior approval of the employer. In circumstances where no unforeseen items of work arise, the contingency sum should remain unspent at practical completion, but in practice, this is a rare (indeed almost unheard of) occurrence. If construction costs are to be controlled successfully, it is essential that there be good communication between all members of the building team. As it falls to quantity surveyors to maintain construction cost records and provide employers with regular financial reviews, it is essential that they attend all meetings when cost matters are discussed and that they receive copies of all documents that may have some bearing on the cost of any project.

It is also essential that the cost effect of any proposed variations, whether emanating from the architect/contract administrator or from any of the other design consultants, is determined as soon as possible so that steps can be taken to minimise its impact on the contract sum. As a precursor to this, the quantity surveyor should be asked to give an early indication of a proposed variation's value before it is issued as it may be that this cost indication results in the abandonment or modification of the proposed variation. However, a contractor's Schedule 2 Quotation means that it is possible to obtain a fixed-price quotation from the contractor for a proposed variation prior to authorising its execution. Whilst using this procedure may take a little longer than the quantity surveyor's estimate, it does provide fixed costs, which should lead to more accurate cost control.

By looking ahead and making early decisions on such matters as the listing of sub-contractors and suppliers and the expenditure of provisional sums, the architect/contract administrator can greatly assist the cost control process.

The regular financial reviews referred to previously will normally be produced by the quantity surveyor and sent to the employer and other members of the design team at monthly intervals, often coinciding with the dates of issue of interim certificates. These reviews should show a forecast of the employer's total financial commitment to the contractor, including the reimbursement of direct loss and/or expense but normally excluding VAT and should, in effect, comprise an estimate of the final adjustment of the contract sum. A list of all the adjustments to be made to the contract sum is set down in SBC clause 4.5, and this can act as the most useful *aide memoire* when preparing a financial review.

A typical financial review is given as Example 29.1.

FUSSEDON KNOWLES & PARTNERS	Financial Review No: 9
Chartered Quantity Surveyors	**Date of Issue:** 12 March 2014
Upper Market Street	**Reference:** 1234
Borchester BC2 1HH	
Works: Shops & Offices, Newbridge Street, Borchester	
Contractor: Leavesden Barnes & Co Ltd	**Employer:** OWGS
Contract Sum: £675,332.00	**Approved Expenditure:** £680,000
Date of Possession: 11 June 2013	**Date for Completion:** 17 April 2015

ESTIMATED CURRENT FINANCIAL COMMITMENT MORE THAN APPROVED EXPENDITURE

	£	£
Contract sum	1,675,332	
Deduct contingencies	1,675,332	
	————	1,655,332
Add estimated value of variations		23,980
		1,679,312
Add/deduct estimated fluctuations in cost of labour and materials		NIL
		1,679,312
Add estimated reimbursement of direct loss and/or expense		2,000
Total estimated final cost		1,681,312
Deduct approved expenditure		1,680,000
OVERSPENT BY		**£ 1,312**

ESTIMATED CURRENT FINANCIAL COMMITMENT LESS THAN APPROVED EXPENDITURE

	£	£
Approved expenditure		
Contract sum		
Deduct contingencies	————	
Add/deduct estimated value of variations	————	
Add/deduct estimated fluctuations in cost of labour and materials	————	
Add estimated reimbursement of direct loss and/or expense	————	
Total estimated final cost		————
UNDERSPENT BY		£————

Notes: All amounts given above are EXCLUSIVE of fees and Value Added Tax

Signature: Date: 12 March 2014

Part IV

Example 29.1   Financial review. Reproduced by permission of Royal Institute of British Architects.

Chapter 30
Interim Payments

Introduction

Prior to 1998, it was a basic principle of English contract law that when a contractor undertook to do work for a fixed sum, he was not due any payment until the whole of the work had been completed, but once the work was complete, he was due full payment of the fixed sum. Whilst this arrangement may have been appropriate for small projects, construction work can involve large sums of money being expended over a number of months or even years. Thus, in order to provide cash flow to the contractor, this basic principle was usually varied by agreement between the parties to allow interim payments to be made as the work proceeded.

The Housing Grants, Construction and Regeneration Act 1996 (the Construction Act) and The Scheme for Construction Contracts (England and Wales) Regulations 1998 (The Scheme) came into effect on 1st May 1998 and changed contract law in respect of what the Construction Act defined as construction contracts, except for those with a residential occupier.

Amongst other things, the Construction Act required that in respect of payments:

- A party to a construction contract is entitled to payment by instalments, stage payments or other periodic payments for any work under the contract unless the duration of the work is less than 45 days. The parties are free to agree on the amounts of the payments and the intervals at which, or the circumstances in which, they become due. In the absence of agreement, such amounts, intervals or circumstances have to be determined by reference to Part II of The Scheme.
- Every construction contract shall provide a mechanism for determining what payments become due under the contract, and when, and shall also provide for a final date for payment of any sum that becomes due. The parties are free

The Aqua Group Guide to Procurement, Tendering and Contract Administration, Second Edition.
Edited by Mark Hackett and Gary Statham.
© 2016 John Wiley & Sons, Ltd. Published 2016 by John Wiley & Sons, Ltd.

to agree on how long the period is to be between the date on which the sum becomes due and the final date for payment. In the absence of such provisions, these matters have to be determined by reference to Part II of The Scheme.

- Every construction contract shall provide for the giving of notice by a party, not later than 5 days after the date on which a payment becomes due from him, specifying the amount of the payment to be made and the basis on which that amount is calculated. In the absence of such provisions, the notice specifying the amount of payment has to be that set down in Part II of The Scheme.
- A party to a construction contract may not withhold payment after the final date for payment unless he has given an effective notice of intention to withhold payment not later than the prescribed period before the final date for payment. The parties are free to agree on the length of the prescribed period before the final date for payment. In the absence of such agreement, the period has to be that set down in Part II of The Scheme.
- Where the sum due under a construction contract is not paid in full by the final date for payment and no effective notice to withhold payment has been given, the person to whom the sum is due has the right to suspend performance of his obligations under the contract.
- A provision making payment under a construction contract conditional on the payer receiving payment from a third party is ineffective, unless that third party is insolvent.

Part 8 of the Local Democracy Economic Development and Construction Act 2009 (DEDC) amends the provisions of the Construction Act. Consequential amendments have also been made to the Scheme by way of the Construction Contracts (England and Wales) Regulations 1998 (Amendment) (England) Regulations 2011 and the Scheme for Construction Contracts (England and Wales) Regulations 1998 (Amendment) (Wales) Regulations 2011. These apply to contracts entered into on or after 1st October 2011.

The main changes to the original provisions regarding payment notices are as follows:

- The previous 'adequate payment mechanism' is now reinforced to forbid the previously allowable pay-when-certified clauses. Such clauses can no longer be used to prevent payment to a sub-contractor on the basis that work is completed by the subcontractor but does not comprise part of a main contractor's certificate. (A 'pay when paid' clause remains permissible only in circumstances where one party becomes insolvent and that party is higher up the contractual chain).
- The contract must specify that the payer, the payee or another specified person will issue the payment notice. A payment notice must be issued, even if the amount of the payment notice is nil. The sum notified in the payment notice becomes 'the notified sum' and becomes payable on or before the final payment date (subject to a valid 'pay-less notice' being served).
- If the relevant party does not issue the payment notice within the permitted time period (not later than 5 days after the payment due date), the other party has a right to issue a payment notice by itself, specifying the amount it believes

is or was due at the payment date. Such sum contained within the payment notice then becomes 'the notified sum'.

■ The paying party must give a 'pay-less notice' if it intends to pay less than 'the notified sum'. The 'pay-less notice' must be given no later than the number of days stated in the contract or, if the contract is silent as to the number of days, 7 days before the final date for payment. A 'pay-less notice' must specify the sum that the paying party considers is due on the day that notice is served and the basis on which that sum is calculated.

In addition to changes to the payment notices, the new provisions amend the rights of suspension for non-payment and under which the suspending party's rights are enhanced. The right to suspend performance now extends to suspending performance in respect of 'any or all' of its obligations. This seeks to strengthen the contractor's position in that it can choose to carry out certain amount of the work or to stop work entirely. The contractor will be entitled to an extension of time if the contractor suspends the performance of the works in its entirety. Such suspension will mean that the contractor will be entitled to receive payment of a reasonable amount in respect of costs and expenses reasonably incurred in exercising its right to suspend performance. This is not only for the period of suspension itself, but also for the period that is a consequence of such suspension.

All standard forms of contract now incorporate such payment provisions as are necessary to comply with current payment legislation. They also provide for money to be held in trust, as retention both during and at the end of the period of construction, primarily to provide a fund from which the employer is able to recover the cost of having defects repaired that the contractor may be unwilling to put right.

When considering the subject of interim payments, it is worth bearing in mind that the contractor has similar obligations to his suppliers and sub-contractors as does the employer to the contractor. Provided that the minimum requirements of the payment legislation are met, interim payment provisions can be in any one of the following forms:

■ payments of pre-determined amounts at regular intervals;
■ payments of pre-determined amounts when the work reaches pre-determined stages of completion; or
■ payments of amounts at regular intervals calculated by a detailed valuation process as the work proceeds.

Payments of pre-determined amounts at regular intervals

This is a method where, for example, a construction project of £1,200,000 is contracted to take 12 months, and it is determined that the contractor will be paid £100,000 per month, adjusted to account for authorised variations, fluctuations, loss and/or expense and retention. The method has the advantage to both parties of minimal administration and cash flow certainty. However, it can lead to problems if the contractor falls behind programme or executes defective work, whereupon over-payment to the contractor for the work done is a probability, which would certainly lead to problems in the event of the contractor going into liquidation before completing the project.

Pre-determined payments at pre-determined stages

This is a method whereby pre-determined amounts, adjusted as mentioned previously, become due for payment upon the proper completion of a pre-determined stage of work. For example, the pre-determined amount of, say, £72,000 (duly adjusted) becomes payable upon the proper completion of all work up to damp-proof course level.

Determining the amounts of interim payments in this way keeps administration costs down and, unlike the previous method, does not lead to overpayment to the contractor if he falls behind programme or if he executes defective work. It can, however, be disadvantageous to the contractor, as it does not take into account the value of unfixed materials and goods either on or off site. Also, the non-completion of a minor, non-critical and low-value element of work in a high-value stage would prevent payment becoming due for the whole stage.

It is interesting to note that the JCT's DB Contract provides for the optional use of this method of payment under clause 4.7 Alternative A – Stage Payments.

Regular payments by detailed valuation

This method of determining the amounts of interim payments is the fairest and most commonly adopted in construction contracts. The implications of defective work, delays, authorised variations, materials on and off site, fluctuations, loss and/or expense and retention can all be readily taken into account in each payment. This procedure does, however, require a regular and substantial input from most members of the building team.

The SBC adopts this method of payment and sets out the procedures to be followed in considerable detail in Section 4 – Payment.

Certificates and payments under the SBC

The obligations of the employer's consultants and the parties to the contract, as set out in Section 4 of the SBC, are briefly as follows.

The architect/contract administrator

The architect/contract administrator has to issue interim certificates stating the amount due to the contractor, to what that amount relates and the basis of calculation of that amount:

- on the dates provided for in the contract particulars up to the date of practical completion or to within 1 month thereafter (these will not be at intervals of less than one calendar month);
- after practical completion as and when further amounts are ascertained as being payable to the contractor; and
- after expiration of the rectification period or upon issue of the certificate of making good, whichever is the later.

Interim certificates are vital to the smooth running of a project, and the architect/contract administrator should bear in mind that:

- certificates must be issued at the appointed time and in accordance with the requirements of the contract;
- he is legally responsible for the accuracy of his certificates; and
- he must remain independent in the issuing of certificates, so as to be fair to both parties.

The quantity surveyor

If contractors submit an application setting out their gross valuation to the quantity surveyor, the quantity surveyor has to make an interim valuation. If the quantity surveyor disagrees with the gross valuation of the contractor, he has to submit a statement to the contractor at the time of making his valuation and in similar detail to the contractor's application, identifying his disagreement. In the absence of an application from the contractor, it is at the architect/contract administrator's discretion whether interim valuations are to be made by the quantity surveyor except where clause 4.21 (choice of fluctuation provisions) applies, which requires the quantity surveyor to make an interim valuation before the issue of each interim certificate.

The employer

Not later than 5 days before the final date for payment of the amount certified due, the employer (or any other person the employer notifies the contractor as being authorised to do so on his behalf) may give written notice to the contractor specifying the sum that he considers to be due and the basis on which that sum has been calculated. This provision enables the employer to levy liquidated damages against the contractor. Such notices are now termed 'pay-less notices' under JCT 2011 and the LDEDC amendments to the Construction Act, but they were previously known as 'withholding notices'. Under the 'withholding notice' provisions, it is arguable that more detail was required to be provided than under the amended legislation; the notice had to include the amount proposed to be withheld or deducted from the due amount, the ground(s) for so doing and the amount attributable to each ground.

If the employer does not give these notices and fails to pay, in full, the amount stated as due in an interim certificate by the final date for payment, that is, within 14 days from the date of issue of the interim certificate, then:

- the contractor is entitled to be paid simple interest on the overdue amount at 5% per annum above the official dealing rate of the Bank of England;
- the contractor is empowered, subject to giving the notice required by clause 4.14, to suspend the performance of his obligations under the contract until payment in full occurs (under JCT 2011 this suspension can be in part only – under JCT 2005 it was all or nothing – as noted earlier in this chapter and in accordance with the LDEDC 2009 amendments to the Construction Act); and
- the contractor is empowered, subject to giving the notices required by clause 8.9, to terminate his employment under the contract and he may also start proceedings in the courts for the recovery of the debt.

Whilst employers have the right, in the circumstances prescribed by and subject to giving the notices required by clause 2.32, to deduct liquidated damages from the amount stated as due in any interim certificate, they do not have the right to interfere with the issue of any architect/contract administrator's certificate. If they do so, it is a matter for which the contractor may determine his employment under the contract. However, contractors would achieve nothing by determining if it were the final certificate that was obstructed or interfered with, and it is doubtful whether they would achieve much by determining their employment under the contract close to or after practical completion. In such circumstances, they would, instead, be better off invoking the procedures for the settlement of disputes or differences prescribed by mediation, adjudication, arbitration or legal proceedings.

When it is recorded in the contract particulars that the employer is a 'contractor', the employer must, when making payment, comply with the provisions of clause 4.7 and must do so in accordance with the Construction Industry Scheme (CIS). Certain non-construction businesses or other concerns are required to act as contractors. These include government departments, some public bodies and businesses that have an average annual expenditure in excess of £1 million on construction operations over 3 years ending with their last accounting date. These organisations are required by HM Revenue and Customs to operate the CIS.

The contractor

Whilst clause 4.11, subject to any agreement between the parties, allows contractors to submit to the quantity surveyor an application setting out their gross valuation pursuant to clause 4.16, they are under no contractual obligation to assist in any way in the preparation of interim valuations or certificates. It is the architect/contract administrator's responsibility to issue the interim certificates at the correct times, and it is the employer's responsibility to make payment to the contractors within the 14-day period.

If the contractor has not submitted an interim application and the architect/contract administrator has not issued an interim certificate within 5 days of a due date, the contractor may issue (at any time thereafter) to the quantity surveyor a (default) interim payment notice in accordance with the provisions of clause 4.11.2.2. Such notice must state the sum it considers to be or have been due at the relevant due date and the basis on which the sum is calculated. In such circumstances the sum stated as due in the contractor's interim payment notice becomes the sum to be paid by the employer under the provisions of clause 4.12.3 (and subject to any valid 'pay-less notice' issued by the employer under 4.12.5). Furthermore, the final date for payment is postponed by the number of days that fell between the last date by which the architect/contract administrator should have issued an interim certificate and the date on which the contractor issued its interim payment notice.

It is also worth noting that although 'pay-less notices' are most likely to be invoked by the employer, the contractor may issue such a notice where the interim payment certificate identifies that a payment is due from the contractor to the employer. In such a scenario, the contractor has the same obligations as the employer to issue a timely notice specifying the sum that it considers to be due and the basis on which that sum has been calculated.

Under the SBC Sub-Contract, not later than 5 days after the date that payment becomes due, the contractor must give written notice to each sub-contractor specifying the amount of the payment in respect of their sub-contract works, to what that amount relates and the basis of calculation of that amount. Not later than 5 days before the final date for interim payment, the contractor may give written notice to each sub-contractor that it intends to pay less than the sum stated in the interim payment notice. Such notices to pay less must specify the amount that the contractor considered to be due at the date the notice is given and the basis on which that sum has been calculated. If the contractor does not issue a valid pay-less notice and fails to pay, in full, the amount stated as due in an interim payment by the final date for payment (i.e. within 21 days after the date on which they become due) then:

- the sub-contractor is entitled to be paid simple interest on the overdue amount at 5% per annum above the official dealing rate of the Bank of England; and
- the sub-contractor is empowered, subject to giving the notice required by clause 4.11 of Standard Building Sub-Contract (SBCSub), to suspend the performance of his obligations under the contract until payment in full occurs.

Interim certificates under the SBC

Whether or not there is a contractual requirement for the quantity surveyor to prepare interim valuations for the purpose of ascertaining the amount to be stated as due in an interim certificate, it is normal practice on most projects for this to be done and for the contractor to co-operate.

According to clause 4.9, the amount to be stated as due in an interim certificate is the gross valuation up to and including a date not more than 7 days before the date of the interim certificate less:

- any amount that may be deducted by the employer as the retention;
- the total amount of any advance payment due for reimbursement to the employer;

- the total amount stated as due in previous interim certificates; and
- any sums paid in respect of an interim payment notice given after the issue of the latest interim certificate (whether adjusted by a 'pay-less notice').

Clause 4.16 provides that the gross valuation shall be the following:

- the total of all the amounts to be included that are subject to retention;
- plus the total of all the amounts to be included that are not subject to retention; and
- less the total of all the amounts to be deducted that are not subject to retention.

The amounts to be included that are subject to retention are the following:

- the total value of the work, properly executed by the contractor, including variations and, where applicable, any adjustment of that value under fluctuation Option C, but excluding any restoration, or repair of loss or damage and removal and disposal of debris treated as a variation following a claim on the insurance policy covering the works;
- the total value of materials and goods delivered to, or adjacent to the works and due to be incorporated, provided they are reasonably and not prematurely delivered – and have been protected from the weather; and
- the total value of any listed items, before their delivery to or adjacent to the works (see *Unfixed materials and goods off site* in the following section).

The amounts to be included that are not subject to retention are the following:

- any amount to be included in interim certificates in respect of
  - statutory fees and charges
  - opening up work for inspection and testing
  - patent rights arising out of compliance with an architect's instruction
  - various insurances under clauses 2.6.2, 6.5, 6.10, B.2.1.2 and C.3.1;
- any amount due to the contractor by way of reimbursement of loss and/or expense arising from matters materially affecting the regular progress, from the discovery of antiquities, or following (non-payment and) suspension of the work (or part thereof) under 4.14.2;
- any amount due to the contractor in respect of any restoration, replacement or repair of loss or damage and disposal of debris that are treated as a variation under paragraphs B.3.5 and C.4.5.2 of Schedule 3 or clause 6.11.5.2; and
- any amount due to the contractor under fluctuation Options A or B.

The amounts to be deducted that are not subject to retention are the following:

- any amount in respect of errors in setting out, which the architect/contract administrator instructs are not to be amended;
- any amount in respect of any defects, shrinkages or other faults, which the architect/contract administrator instructs are not to be made good;

- any amount in respect of work completed by another person following non-compliance with an architect/contract administrator's instruction in accordance with clause 3.11;
- any amount in respect of work not in accordance with the contract, which the architect/contract administrator may allow to remain; and
- any amount allowable by the contractor to the employer under fluctuations Options A or B, if applicable.

Unfixed materials and goods on site

The value of all materials and goods stored on site must be included in the valuation, provided that they are adequately protected and have not been brought to site prematurely. As an extreme example, unless there was some prior agreement on the matter, the architect/contract administrator is entitled to withhold payment for items of furniture brought onto site whilst only work on the foundations is in progress. Any materials or goods included in the amount stated as being due in an interim certificate that has been paid by the employer become the property of the employer, and although the contractor remains responsible for their loss or damage, they must not be removed from the site.

Unfixed materials and goods off site

Only the value of materials or goods stored off site that have been listed by the employer in a list supplied to the contractor and annexed to the contract documents can be considered for inclusion in a gross valuation, and then only if the following criteria are met:

- the contractor has provided the architect/contract administrator with reasonable proof that the property in the listed items is vested in the contractor;
- if so stated in the contract particulars, the contractor has provided a bond in favour of the employer from a surety approved by the employer in respect of payment for the listed items;
- the listed items are in accordance with the contract;
- the listed items are set apart or are clearly and visibly marked, and identify the employer and to whose order they are held and their destination as the Works; and
- the contractor has provided the employer with reasonable proof that the listed items are insured against loss or damage for their full value.

Any listed items included in the amount stated as being due in an interim certificate that has been paid by the employer become the property of the employer, and whilst the contractor remains responsible for their loss or damage, and the cost of their storage, handling and insurance, they must not be removed from the premises where they are stored except for the purpose of their delivery to site.

Retention under the SBC

The contract provides that, in the calculation of the amount to be stated as due in an interim certificate, the employer may deduct 'the retention' from the gross valuation. The total amount of the retention at any one time is ascertained by application of the rules given in clause 4.20 and comprises the following:

- the total value of work (or sections) properly executed that has not reached practical completion and of site materials and listed items included in the gross valuation – at the retention percentage; and
- the total value of work (or sections) that has reached practical completion or has been taken into the employer's possession by agreement with the contractor, but for which a certificate of making good defects has not been issued – at half the retention percentage.

The retention percentage is 3% unless a lower rate is specified in the appendix to the contract. It is to be noted that no retention is held against the value of work for which a certificate of making good defects has been issued.

The treatment of the retention is subject to the rules given in clause 4.18, which provide as follows:

- the employer's interest in the retention is fiduciary as trustee for the contractor, but without an obligation to invest (in other words, it is money held by the employer in trust for the contractor);
- the architect/contract administrator has to prepare, or instruct the quantity surveyor to prepare, a statement of the contractor's retention at the date of each interim certificate and copies are to be issued to both the employer and the contractor; and
- the employer shall, at the request of the contractor, place the retention in a separate, appropriately designated banking account and certify to the architect/contract administrator, with a copy to the contractor, that this has been done (this rule does not apply when a local authority is the employer).

Payments to sub-contractors under the SBC

The SBC contains very few references to a contractor's obligation to pay sub-contractors, which are obviously dealt with in each respective sub-contract. However, under clause 3.9.2, it is stated that each sub-contract must provide that if the contractor fails to pay properly any amount due to the sub-contractor by the final date for payment, the contractor shall pay interest on the amount not properly paid. The interest that the sub-contractor is entitled to receive is simple interest at the rate of 5% above the official dealing rate of the Bank of England.

Value added tax

Clause 4.6 makes it quite clear that the contract sum is exclusive of value added tax. The responsibility for the payment of this tax lies with contractors, and they in turn will invoice employers appropriately. If, after the base date of the contract, the supply of goods and services to the employer becomes exempt from VAT (value added tax), because of a change in the tax legislation, the employer remains responsible for the payment of input tax to the contractor, that is, the amount that the contractor has paid to its sub-contractor/suppliers for the supply of goods or services and that would not otherwise be recovered by the contractor due to the employer's exemption.

Valuation and certificate forms

Except where employers have their own forms on which they require interim payments to be certified, it is normal practice for standard forms published by the professional bodies to be used. Valuation forms are published by the RICS for the use of quantity surveyors. These comprise the valuation shown in Example 30.1 and the statement of retention (which accompanies the valuation form as an appendix) shown in Example 30.2. RIBA Publications publishes forms for architects to use in connection with interim certificates. These are the interim certificate and the statement of retention. This can be filled in directly from the information given in the quantity surveyor's valuation, and a completed specimen is shown in Example 30.3. All these forms are published in separate pads, and each pad contains notes on their use. It will also be seen that the quantity surveyor's valuation form contains several notes that form a useful reminder of the contractual obligations.

Valuation for JCT Standard Building Contract (2011 Edition)

Surveyor Fussdean Knowles & Partners
Upper Market Street
Borchester BC9 1HH

Works   New School Library
Pack Street
Packbridge

Valuation no: 8
Date of issue: 20 September 2014
Reference: RWP/03102

To Architect / Contract Administrator
Ivor Bach Associates
Prospects Drive
Fanbridge

*I / we have made, in accordance with the terms of the
Contract, an Interim Valuation, the basis on which the amount shown as due has been calculated is
clause 4.9.2 of the Conditions of Contract, and report as follows:

As at   15 September 2014

Gross valuation
(excluding any work or material notified to me/us by the Architect /Contract Adminstrator in writing as
not being in accordance with the Contract)   **£347,260.00**

Less total amount of retention, as attached Statement   **£17,362.00**

Employer
Cosmeston Preparatory School
Fanbridge

Less total amount of Interim Certificates previously issued by the Architect/Contract Administrator
up to and including Interim Certificate No. 7
and any advance payment (if any) due for reimbursement by the date of the next Certificate.   **£258,133.00**

Balance   Seventy one thousand, seven hundred and sixty five pounds   **£71,765.00**

Contractor
L&M Construction Ltd
Perry Road
Fanbridge

Signature   *R.W.Pipe*

Surveyor *FRICS/ARICS/*FRICS

Contract sum £506,207.00

Notes:
1. All the above amounts are exclusive of VAT.
2. The balance stated is subject to any statutory deductions which the Employer may be obliged to make under the provisions of the
Construction Industry Scheme where the Employer is classed as a 'Contractor' for the purposes of the relevant Act.
3. It is assumed that the Architect / Contract Administrator will satisfy him or herself that there is no further work or material which is not in accordance with the Contract.

* *Delete as appropriate*

© RICS 2011

Example 30.1   Interim valuation.

Part IV

Part IV

Statement of Retention Values

RICS © RICS 2011

Surveyor Fussedean Knowledes & Partners
Upper Market Street
Borchester, BC2 1HH

Works New School Library
Pack Street
Fairbridge

This statement relates to:
Valuation No: 8
Date of Issue: 20/09/2014
Reference: RWP/03102

Description of Works:	Gross Valuation	Basis of Gross Valuations (see note 1) Clause No.	Amount Subject to (see note 2):			Amount of Retention	Net Valuation	Amount Previously notified	Balance
			Full Retention of %	Half Retention of %	No Retention				
Classroom Extension	£347,260.00		£347,260.00	-	-	£17,362.00	£329,898.00	£258,133.00	71,165.00
	£347,260.00		£347,260.00			£17,362.00	£329,898.00	£258,133.00	£71,165.00

Notes: 1. The sums stated are exclusive of VAT.
2. See clause 4.13/4.14 for rules for ascertainment of retention.

Example 30.2 Statement of retention.

Interim Certificate

SBC

Issued by: Ivor Barch Associates
address: Prospects Drive, Fairbridge

Employer: Cosmeston Preparatory School
address: Fairbridge

Contractor: L&M Construction Ltd
address: Ferry Road, Fairbridge

Works: New School Library
situated at: Park Street, Fairbridge

Contract dated: 14 December 2013

Job reference: IBA/94/20

Certificate no: 8

Date of valuation: 15 September 2014

Due date: 18 September 2014

Date of issue: 23 September 2014

Final date for payment: 2 October 2014

This Interim Certificate is issued under the terms of the above-mentioned Contract.

Gross Valuation (calculation attached)	£	347,260.00
Less Retention as detailed on the attached Statement of Retention	£	17,363.00
Sub-total	£	329,897.00
Less reimbursement of advance payment (statement attached)	£	-
Sub-total	£	329.897.00
Less total amount previously certified	£	258,133.00
Sub-total	£	71,764.00
Less payments referred to in clause 4.9.2.4	£	-
Net amount for payment	**£**	71,764.00

All amounts are exclusive of VAT. The Employer shall in addition pay the amount of VAT properly chargeable.

I/We hereby certify that the **amount due** to the Contractor from the Employer is (in words)

Seventy one thousand, seven hundred and sixty pounds

To be signed by or for the issuer named above

Signed _____ R.w.P.px

This is not a Tax Invoice.

Distribution	☐ Employer	☐ Contractor	☐ Quantity Surveyor	☐ File copy

for SBC   CONTRACT ADMINISTRATION FORMS   © RIBA Publishing 2011

Part IV

Example 30.3 Interim certificate.

Chapter 31
Completion, Defects and the Final Account

A building contract does not come to an end until the architect/contract administrator issues his final certificate, and even then, actions for breach of contract can be commenced within 6 or 12 years of the breach, depending on whether the contract had been executed under hand or as a deed (see Chapter 25). Completion of the works can occur in three instances as follows:

- practical completion;
- sectional completion (if applicable) or partial possession; and
- completion of making good defects.

Practical completion

In the SBC, practical completion of the works or section is defined by reference to clause 2.30. When in the opinion of the architect/contract administrator the contractor:

- has achieved practical completion of the works or section, including all authorised variations;
- has provided such information as is reasonably required by the principal designer for the preparation of the health and safety file in accordance with the CDM regulations; and
- has supplied the employer with the required drawings and information, including any contractor's design documents, showing the building's maintenance and operation,

then the architect/contract administrator is required to issue a certificate to that effect. Practical completion of the works or section will, for the purposes of the

The Aqua Group Guide to Procurement, Tendering and Contract Administration, Second Edition.
Edited by Mark Hackett and Gary Statham.
© 2016 John Wiley & Sons, Ltd. Published 2016 by John Wiley & Sons, Ltd.

contract, be deemed to have taken place on the date stated in that certificate. This date is significant as it has an important influence on a number of conditions in the contract and any disputes that may arise under it. RIBA Publications publishes a standard pro forma Certificate of Practical Completion, and a completed copy of this form is given as Example 31.1 at the end of this chapter.

Over the years, the courts have defined practical completion of the works in differing ways, but the safest approach for any architect/contract administrator to adopt is to certify practical completion of the works or section only when every item of work has been satisfactorily completed. However, the architect/contract administrator is often put under pressure to certify practical completion prematurely to enable the employer to gain occupation of his building. The consequences to the employer and the contractor of issuing the certificate of practical completion whilst items of work remain incomplete, even if they are of a minor nature, must be considered by the architect/contract administrator – for example, the contractor will have to complete work in an occupied (and perhaps otherwise finished) building, and this may cause additional problems related to health, safety, security and damage to finished works. If an architect/contract administrator were minded to issue the certificate of practical completion when items of work remain incomplete, he would be well advised separately to identify, in writing to the contractor, the items of work that remain to be completed, an order of priority for their completion and the date(s) by when they are to be completed. Practical completion ought not to be certified when there is any outstanding *defective* work (as distinct from incomplete work).

The achievement of practical completion is used in the SBC as the datum for the effective commencement or cessation of various provisions. These are as follows:

Clause	Description
2.4	The contractor retains possession of the site and the works or section up to and including the date of issue of the certificate of practical completion, and the employer is not entitled to take possession of the works or section until that date, except as provided for in clause 2.33 (see later in this chapter).
2.28.5	Not later than the expiry of 12 weeks after the date of practical completion, the architect/contract administrator has to fix a completion date later or earlier than previously fixed or confirm the completion date previously fixed. Nevertheless, an architect/contract administrator cannot fix an earlier completion date than the original contract completion date. In re-fixing a completion date, he cannot alter the length of any pre-agreed adjustment except in the case of a Schedule 2 Quotation where the relevant Variation is itself the subject of a relevant omission.
2.32	The employer is entitled to recover liquidated damages from the contractor in respect of the period between the completion date and the date of practical completion if a non-completion certificate has been issued in accordance with clause 2.31.
2.38	The rectification period (6 months unless the contract particulars state to the contrary) runs from the day named in the certificate of practical completion of the works or section.
2.38	Defects, shrinkages or other faults due to materials or workmanship not being in accordance with the contract, or failure by the contractor

to comply with the contractor's designed portion, which occur following practical completion of the works or section, have to be made good by the contractor at no cost to the employer.

2.40   The contractor has to supply the employer with the required drawings and information showing and/or describing any contractor's designed portion work as built and, where relevant, its maintenance and operation, before the date of practical completion.

4.5   Not later than 6 months after issue of the practical completion certificate, the contractor has to provide to the architect/contract administrator or to the quantity surveyor all documents necessary for the purposes of the adjustment of the contract sum.

4.9.1   Interim certificates have to be issued on the dates provided in the contract particulars up to the date of practical completion or to within 1 month after it. Thereafter, interim certificates have to be issued at intervals of 2 months (unless otherwise agreed) up until the issue of certificate of making good or the end of the rectification period (whichever is later).

4.20.3   In interim certificates, the full retention percentage may be deducted from the value of work that has not reached practical completion, but only half the retention percentage may be deducted from the value of work that has reached practical completion (until such time as the certificate of making good defects is issued under clause 2.39).

6.5.1.5   Where any part of the works or sections is not subject of a certificate of practical completion, injury or damage to the works is excluded from any insurance required pursuant to clauses 6.5.1.1–6.5.1.9 (insurance – liability, etc. of employer).

6.7   The joint names insurance policies required by the applicable clause(s) (within Schedule 3 clauses A.1, B.1 and C.1) have to be maintained by either the contractor or the employer, as the case may be, up to and including the date of issue of the certificate of practical completion (or the date of termination of the contractor's employment if that is earlier). Upon deemed practical completion of a relevant part, in the case of vacant possession, the employer is responsible for insuring the completed part of the works.

7.2   Where the contract particulars state that this clause applies the employer may, at any time after practical completion of the works or section, assign to a transferee or lessee the right to bring proceedings in the name of the employer to enforce any of the terms of the contract made for the benefit of the employer.

8.4   If the contractor makes a specified default prior to the date of practical completion, then, subject to giving the appropriate notices, the employer is entitled to terminate the employment of the contractor.

8.9   If the carrying out of the works, or substantially the whole of the uncompleted works, is suspended for a continuous period (of a length stated in the contract particulars) or the employer makes a specified default, prior to the date of practical completion, then, subject to giving the appropriate notices, the contractor is entitled to terminate his employment.

8.11   If the carrying out of the works, or substantially the whole of the uncompleted works, is suspended for a continuous period (of a length stated in the contract particulars), prior to the date of practical completion, for one or more of the specified reasons, then, subject to giving the appropriate notices, either party is entitled to terminate the contractor's employment.

If the project is large, it is quite feasible to have the situation where a certificate of practical completion is issued in respect of one section before work commences on other sections. Similarly, the certificate of making good defects can be issued in respect of one section, whilst the work remains incomplete on other sections.

Partial possession

As previously noted, sectional completion arises when the employer requires the works to be completed in phased sections. Partial possession, however, arises out of a contract agreement between the employer and the contractor as provided for in SBC clauses 2.33 to 2.37. If, before practical completion, the employer, with the consent of the contractor, takes possession of a part of the works or a section, clauses 2.33 to 2.37 apply with the following effects:

- The architect/contract administrator has to issue immediately, to the contractor, a written statement identifying the part of the works or section taken into possession (the relevant part) and the date on which that possession occurred (the relevant date).
- In respect of the relevant part, practical completion is deemed to have occurred and the rectification period is deemed to have commenced on the relevant date.
- The architect/contract administrator has to issue a certificate of making good defects in respect of the relevant part when any defects, shrinkages or other faults in the relevant part that were required to be made good have been made good.
- The contractor and the employer are relieved of their respective duties to insure the relevant part under clause 6.7 and Option A, Option B or Option C, whichever is applicable, but when Option C is applicable, the employer will be obliged to insure the relevant part from the relevant date.
- The rate of liquidated damages (LD) stated in the contract particulars has to be proportionally revised as follows:

Revised LD

$$= \frac{\text{Original LD} \times (\text{Contract sum} - \text{Value of Relevant Part in Contract sum})}{\text{Contract sum}}$$

Possession of the building

A short time before practical completion, the architect/contract administrator should ensure that the contractor has removed all plant, surplus materials, rubbish and temporary works from the site. When practical completion has been achieved and the architect/contract administrator has issued the certificate to that effect, the contractor ceases to be responsible for the works or section of the site and relinquishes possession of them to the employer. At this time, the employer becomes responsible for all insurance matters relating to the site, the building(s) and contents.

At this stage, it is advisable for the design team and the contractor to have a "hand-over" meeting with the employer (see also Chapter 26 on meetings). During this meeting:

- the contractor can formally hand over to the employer all keys, properly labelled;
- the contractor can hand over to the employer the building log-book, referred to in the Building Regulations 2006 Part L, giving details of the installed building services plant and controls, the method of operation and maintenance, and other details that collectively enable energy consumption to be monitored and controlled;
- the employer and/or his staff can be fully briefed on the operation of the building and its services (when the building and/or its services are complex, this exercise may involve special training sessions for those who are to operate and maintain the facilities);
- the contractor can hand over to the employer all relevant operating and maintenance manuals, guarantees and, where appropriate, a supply of spare parts for maintenance of critical equipment and so on;
- the design team and/or the contractor can hand over to the employer a full set of "as built" drawings; and
- the principal designer can hand over to the employer the completed health and safety file.

Defects and making good

SBC clause 2.38 requires that, not later than 14 days after the expiry of the rectification period, the architect/contract administrator has to prepare and deliver to the contractor as an instruction a schedule of all defects, shrinkages or other faults that are due to materials or workmanship not in accordance with the contract that have appeared during the rectification period (this was previously referred to as the defects liability period).

In practice, it is usual for the architect/contract administrator to keep a record of such defects as they occur or are noticed and for this record to form the basis of his schedule, which would normally be compiled by the design team during the course of the works followed by a thorough inspection at the end of the rectification period. Under normal circumstances, the contractor has to make good all items on the schedule of defects at no cost to the employer and within a reasonable time of having received the instruction. However, SBC clause 2.38 does allow the architect/contract administrator, with the employer's consent, to instruct the contractor not to make good some or all of the scheduled items. When the architect/contract administrator does so instruct, an appropriate deduction is made from the contract sum in respect of those items not to be made good.

If the architect/contract administrator requires any defect, shrinkage or fault to be made good during the rectification, SBC clause 2.38 permits him to so instruct the contractor, who has to comply with the instruction within a reasonable time. If the contractor fails to make good any item within a reasonable period, the architect/contract administrator may, after having given the contractor the notice required by SBC clause 3.11, employ and pay others to execute the work and deduct all costs thereby incurred from monies due to the contractor. If there is insufficient money due to the contractor fully to defray the costs incurred, the shortfall is recoverable by the employer from the contractor as a debt.

When the architect/contract administrator is satisfied that all defects, shrinkages and other faults that were formally required to be made good have been made good, a certificate to that effect must be issued. RIBA Publications has produced a pro forma certificate of completion of making good defects, and a completed copy is given as Example 31.2 at the end of this chapter.

Final account

The responsibilities of the contractor, the architect/contract administrator and the quantity surveyor in connection with the final account are set out at the beginning of SBC clause 4.5 and which provides that:

- not later than 6 months after practical completion of the works, the contractor has to send to the architect/contract administrator (or the quantity surveyor if so instructed by the architect/contract administrator) all documents necessary for the adjustment of the contract sum;
- within 3 months of the contractor sending all necessary documents, the architect/contract administrator (or the quantity surveyor if so instructed by the architect/contract administrator) has to ascertain any outstanding loss and/or expense, and the quantity surveyor has to prepare a statement of all other adjustments to be made to the contract sum (the ascertainment of loss and/or expense and the statement of adjustments are, together, commonly referred to as the final account); and
- when complete, the architect/contract administrator has to send a copy of the final account to the contractor.

It is worth noting that, under the SBC, the architect/contract administrator and the quantity surveyor have unilateral responsibility for the preparation of the final account – there is no requirement for them to obtain the contractor's agreement to their ascertainment and adjustments. However, in practice, it is normal for the design team and the contractor to co-operate in reaching an agreed final account. When such agreement cannot be reached the contractor's only recourse is to invoke the contract dispute resolution procedures.

Adjustment of the contract sum

SBC clause 4.3 also sets out all of the matters that must be dealt with in the final account in order properly to adjust the contract sum in accordance with the conditions of contract. These are summarised in the following section.

Part IV

To be deducted or added as the case may be

- the amount of any valuation agreed by the employer and the contractor for a variation;
- the value of any Schedule 2 Quotation for a variation or for acceleration and for which the architect/contract administrator has issued a confirmed acceptance; and
- the value of any variation in the premium for renewing terrorism cover under the joint names policy, when the contractor insures the works under insurance Option A under Pool Re Cover.

To be deducted

- all provisional sums and the value of all work included by way of approximate quantities in the contract bills;
- the amount of the valuation of variations, which are omissions;
- the amount included in the contract bills for work, which, due to a variation, has to be executed under substantially changed conditions;
- the amount included in the CDP analysis, which, due to a variation, has been omitted from the employer's requirements or has to be executed under substantially changed conditions;
- any amount in respect of errors in setting out, which the architect/contract administrator instructs are not to be amended;
- any amount in respect of work not in accordance with the contract, which the architect/contract administrator allows to remain;
- any amount in respect of any defects, shrinkages or other faults, which the architect/contract administrator instructs are not to be made good;
- any amount allowable to the employer for whichever fluctuation option applies;
- any other amount, which is required, by the contract, to be deducted from the contract sum; and
- an appropriate amount of additional costs incurred by the employer through the employment of another party or parties to carry out works following the contractor's non-compliance with an instruction or refusal by contractor to remedy breaches of the Joint Fire Code.

To be added

- any amount payable by the employer in respect of statutory fees and charges legally demandable under any of the statutory requirements;
- any amount payable by the employer for royalties arising out of compliance with an architect/contract administrator's instruction;
- any amount payable by the employer for opening up work for inspection and testing;
- any amount payable by the employer to the contractor for complying with an architect/contract administrator's instruction to take out a policy of insurance under clause 6.5.1;
- the amount of the valuation of variations, other than those that are omissions;

- the amount of the valuation of any work, including CDP work, which due to a variation has to be executed under substantially changed conditions;
- the amount of the valuation of work executed as the result of instructions relating to the expenditure of provisional sums and all work included by way of approximate quantities in the contract bills;
- any amount ascertained in respect of loss and/or expense arising from relevant matters materially affecting the contractor's regular progress (of the works);
- the amount of any costs incurred by the contractor if the employer defaults under insurance Option B or Option C, or under Option A where an additional premium is required by the insurers due to the employer's early use of the site, works or part of them under clause 2.6.1;
- the initial and renewal costs of terrorism cover where Option A applies and cover other than Pool Re Cover is required;
- the net additional cost of alternative or additional terrorism cover where terrorism cover becomes unavailable or is only available at a reduced scope or level and the contractor has been instructed to provide such cover;
- the costs of compliance with revisions to the Joint Fire Code made after the Base Date recorded in the contract where it is stated in the contract that such costs are to be borne by the employer;
- any amount payable to the contractor for whichever fluctuation option applies;
- any other amount that is required, by the contract, to be added to the contract sum; and
- costs and expenses reasonably incurred by the contractor as a result of its suspension of the works following failure by the employer to make interim payment in accordance with the requirements of the contract.

It is good practice for each of the foregoing items to be identified separately in the final account and for separate amounts to be given for each variation.

Practical considerations

Whenever measurements have to be taken for the purpose of valuation, the contractor must be given the opportunity of being present and of taking whatever notes and measurements he may require.

It is a worthwhile exercise to draft the final account as construction proceeds, as it can be of considerable use to the building team for the purposes of:

- post-contract cost control and the preparation of regular financial reviews;
- valuations for interim certificates; and
- speeding up the production of the final account once practical completion of the works has been achieved.

Whether this is done, the design team must bear in mind the SBC requirement to complete the final account within 3 months of receipt from the contractor of all the documents necessary for its production. Any delays in doing so are likely to cause either the contractor or the employer to incur additional financing costs.

Part IV

Finally, clause 4.2 states that any adjustments to the contract sum shall only be carried out in accordance with the express provisions of the contract. Despite this requirement and the architect/contract administrator's unilateral responsibility, or the quantity surveyor's if so instructed, it is the adjustment of the contract sum and the subsequent final account statement that is frequently disputed by contractors. In some instances, and contrary to clause 4.2, the employer and contractor may pragmatically choose to agree a "commercial" settlement. Should a commercial settlement be negotiated and agreed on, this should be recorded in a separate agreement, which is itself a contract.

Final certificate

SBC clause 4.15 requires the architect/contract administrator to issue the final certificate within 2 months of whichever of the following events occurs last:

- the end of the rectification period in respect of the works or, where there are sections, expiry of the period for the last section; or
- the date of issue of the certificate of making good in respect of the works or, where there are sections, issue of the last certificate of making good; or
- the date on which the architect/contract administrator sends a copy of the final account to the contractor.

The amount of the balance due from either the employer to the contractor or vice-versa expressed in the final certificate is the difference between the total of the final account and the amounts previously stated as due in interim certificates plus the amount of any advance payments paid. The same clause also preserves the contractor's rights to any amounts previously included in interim certificates that have not been paid by the employer. RIBA Publications publishes a pro forma final certificate– a completed copy is given as Example 31.3 at the end of this chapter.

SBC clause 1.10 prescribes the intended effect of the final certificate (if not contested), which can be summarised as providing conclusive evidence that:

- where the particular qualities of any materials and so on or standard of any workmanship is to be for the approval of the architect/contract administrator, such items are to the reasonable satisfaction of the architect/contract administrator;
- save for accidental errors, the contract sum has been properly adjusted in accordance with all the terms of the contract;
- only those extensions of time that are due under the contract have been given; and
- the reimbursement of loss and/or expense is in full and final settlement of all and any claims that the contractor may have in respect of matters affecting the regular progress, whether arising from breach of contract, duty of care, statutory duty or otherwise.

Part IV

Issued by: Ivor Barch Associates
address: Prospects Drive, Fairbridge

Practical Completion Certificate

SBC / IC / ICD / MW / MWD

Employer: Cosmeston Preparatory School
address: Fairbridge

Contractor: L&M Construction Ltd
address: Ferry Road, Fairbridge

Job reference: IBA/94/20

Certificate no: 1

Issue date: 28 January 2015

Works: New School Library
situated at: Park Street, Fairbridge

Contract dated: 14 December 2013

Under the terms of the above-mentioned Contract,

I/we hereby certify that in my/our opinion

practical completion of the Works has been achieved

*Delete if not applicable

* and the Contractor has supplied the specified documents and drawings relating to the Contractor's Designed Portion

* and the Contractor has complied with the contractual requirements in respect of information for the health and safety file

on   28 January   20 15

To be signed by or for the issuer named above

Signed _RWBar_

Distribution				
☐ Employer	☐ Structural Engineer	☐ CDM Co-ordinator	☐	
☐ Contractor	☐ M&E Consultant	☐	☐	
☐ Quantity Surveyor	☐ Clerk of Works	☐	☐ File	

for SBC / IC / ICD / MW / MWD   CONTRACT ADMINISTRATION FORMS   © RIBA Publishing 2011

Part IV

Example 31.1   Certificate of practical completion.

Certificate of

Making Good

SBC/IC/ICD

Issued by: Ivor Barch Associates
address: Prospects Drive, Fairbridge

Employer: Cosmeston Preparatory School
address: Fairbridge

Job reference: IBA/94/20

Contractor: L&M Construction Ltd
address: Ferry Road, Fairbridge

Certificate no: 1

Issue date: 28 August 2015

Works: New School Library
situated at: Park Street, Fairbridge

Contract dated: 14 December 2013

Under the terms of the above-mentioned Contract,

I/we hereby certify that the Contractor's obligation to make good any defects, shrinkages or other faults which have appeared during the Rectification Period and been notified to the Contractor

relating to

*Delete as appropriate

* the Works referred to in the Practical Completion Certificate

no.   1   dated   28 January 2015

* Section no. of the Works referred to in the

Section Completion Certificate

no. dated

* the part of the Works identified in the Notice of Partial Possession by the Employer

no. dated

have in my/our opinion been discharged on

28 August   20 15

To be signed by or for the issuer named above

Signed *R.W.Pipe*

Distribution				
☐ Employer	☐ Structural Engineer	☐ CDM Co-ordinator	☐	
☐ Contractor	☐ M&E Consultant	☐	☐	
☐ Quantity Surveyor	☐ Clerk of Works	☐	☐ File	

for SBC/IC/ICD   CONTRACT ADMINISTRATION FORMS   © RIBA Publishing 2011

Example 31.2   Certificate of making good defects.

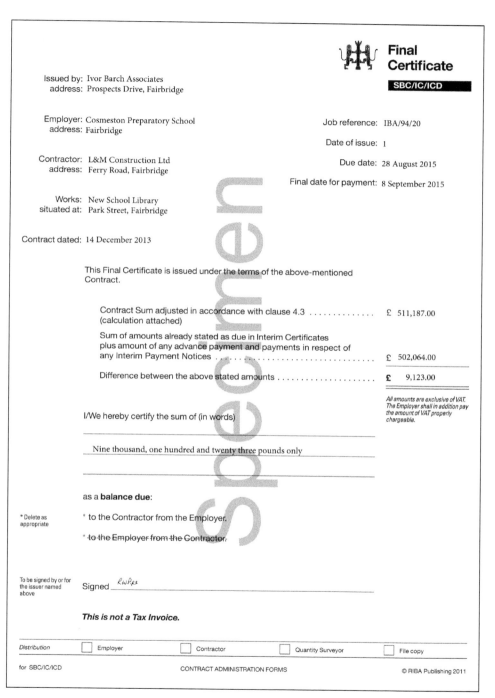

Example 31.3   Final certificate.

Chapter 32
Delays and Disputes

Introduction

Delays to the progress of the works and the loss and/or expense associated with such delays are perhaps the most common causes of dispute encountered in the administration of building contracts. It is not surprising, therefore, that the draughtsmen of the SBC paid particular attention to the procedures to be followed when delays occur, or are foreseen, as a result of which the contract period might become extended and/or the contractor may become entitled to the reimbursement of loss and/or expense.

If the employer (or those for whom he is responsible), through default, prevents the contractor from completing the works by the contracted completion date and there are no express contractual provisions enabling the employer to set a new date for completion, the contractor's obligation is transformed from one of having to complete by the contracted date to one of completing in a reasonable time. In other words, time is 'at large' and employers lose any right that they may otherwise have to levy liquidated damages. It may be seen, therefore, that far from favouring only contractors, the existence and proper execution of an extension of time clause within a contract is of the utmost importance to employers in that it preserves their other contractual rights should they or their representatives cause the works to be delayed by default.

All delays will fall into one of the following three categories:

- delays caused by the contractor;
- delays caused by the employer or his representatives (for whom he is responsible); or
- delays caused by events outside the control of both the contractor and the employer (often referred to in the industry as 'shared risk events' or 'neutral events').

The Aqua Group Guide to Procurement, Tendering and Contract Administration, Second Edition.
Edited by Mark Hackett and Gary Statham.
© 2016 John Wiley & Sons, Ltd. Published 2016 by John Wiley & Sons, Ltd.

Only delays caused by employers or their representatives or by neutral events are recognised by SBC clause 2.29 as relevant events that may give rise to an extension of time; clause 2.27.1 requires the contractor to give written notice to the architect/contract administrator whenever it becomes reasonably apparent that the progress of the works or any section is being or is likely to be delayed, irrespective of cause.

Delays caused by the contractor

Within this category are all those delays that could be avoided if the contractor proceeds regularly and diligently with the works and uses best endeavours at all times to prevent delays. The SBC effectively requires contractors to take all measures within their control to ensure:

- that there is an adequate labour force on the job;
- that the necessary goods and materials are on site whenever they are needed; and
- that the works are not delayed by sub-contractors.

If the contractor does not take all such measures and delays occur, there is no provision in the SBC to enable the architect/contract administrator to grant an extension of time in respect of them. The responsibility for any such delays then lies with the contractor, and under SBC clause 2.32, he may well become liable to the employer for liquidated damages in respect of the delay period.

If the contractor fails to proceed regularly and diligently with the works, the employer's ultimate sanction, subject to giving the required notices, is to terminate the contractor's employment under SBC clause 8.4.

Delays caused by the employer or his representatives

Certain of the relevant events recognised by SBC clause 2.29 result from action or inaction on the part of the employer or his representatives. These are as follows:

- compliance with the architect/contract administrator's instructions requiring a variation or other matters or instructions that are to be treated as a variation under the conditions of the contract;
- compliance with the architect/contract administrator's instructions to open up work for inspection or testing unless the test or inspection shows that the work is not in accordance with the contract;
- compliance with the architect/contract administrator's instructions in regard to any discrepancies within the contract documents;
- compliance with the architect/contract administrator's instructions in regard to the postponement of any work; and
- compliance with the architect/contract administrator's instructions in regard to the expenditure of undefined provisional sums;

Part IV

- the deferment by the employer of giving possession of the site or any section when clause 2.5 applies;
- compliance with clause 3.22.1 and an architect/contract administrator's instructions regarding antiquities;
- the execution of work for which an approximate quantity is included in the contract bills, which is not a reasonably accurate forecast of the quantity of work required;
- suspension by the contractor of the performance of his obligations under the contract as a result of the employer's failure to pay a due amount in full; and
- any impediment, prevention or default, whether by act or omission by the employer, the architect/contract administrator, the quantity surveyor or any of the employer's personnel except to the extent that it was caused or contributed to by any default, whether act or omission, of the contractor or any of the contractor's personnel.

The final bullet point is a catch-all 'relevant event'. A significant number of the employer's obligations under the contract are carried out by the design team, and many of these involve assisting contractors in complying with their own obligations. As an example, the architect/contract administrator's failure to issue design information in accordance with the information release schedule would fall within 'any impediment, prevention or default' by the architect/contract administrator for whom the employer is responsible.

Not only do all delays within this category entitle the contractor to an extension of time if they are likely to delay the completion of the works beyond the completion date, but they may also cause the contractor to suffer loss and/or expense, which, if incurred and subject to the issue of the relevant notices, is reimbursable under the provisions of SBC clause 4.23.

Delays caused by events outside the control of either party

There are occasions when delays arise due to circumstances over which neither the employer nor the contractor has any control. These neutral events, which are also recognised by SBC clause 2.29, deal with delays arising because of the following:

- a statutory undertaker executing, or failing to execute, work in pursuance of its statutory obligations in relation to the works;
- exceptionally adverse weather conditions;
- loss or damage caused by fire, lightning, explosion, storm, flood, escape of water from any water tank, apparatus or pipe, earthquake, aircraft and other aerial devices or articles dropped therefrom, riot and civil commotion but excluding excepted risks;
- civil commotion or the use or threat of terrorism and/or the activity of the relevant authority in dealing with such use or threat;
- strike, lockout or local combination of workmen affecting any of the trades employed on the works or any of the trades engaged in the preparation,

manufacture or transportation of any of the goods or materials required for the works, or persons engaged in the design for the contractor's design portion;
- the ceasing of work and receipt of further instructions from the architect/contract administrator to prevent endangering any fossils, antiquities and other objects of interest or value that may be found on the site;
- exercise by the UK government of any statutory power that directly affects the execution of the works; and
- force majeure.

With one exception to which we shall return, all delays within this category entitle the contractor to an extension of time if they are likely to delay the completion of the works beyond the completion date, but they do not entitle the contractor to reimbursement of any loss and/or expense incurred. In this way, the parties share the burden of any delays caused by these neutral events because the contractor is relieved of the liability to pay or allow liquidated damages, yet he is not entitled to recover any loss and/or expense that he might have incurred.

The exception referred to concerns delays caused by compliance with the architect/contract administrator's instructions in regard to what is to be done with any fossils, antiquities and other objects of interest or value that may be found on the site. In such an instance, the contractor may also recover loss and/or expense.

It is to be noted that in the case of delays, the SBC imposes an obligation on the contractor to take all practicable steps to avoid or reduce the delay. Should such delays occur, the architect/contract administrator needs to be satisfied that the contractor has taken all such steps before consideration is given to granting an extension of time.

Differences in the interpretation of certain of these neutral relevant events may give rise to disputes. It is therefore worth considering some of them in more detail.

Force majeure

This phrase comes from French law (Code Napoléon) and does not have a precise definition in English. Its literal translation is 'superior force', and in its broadest usage, examples would include acts of God, war, strikes, weather conditions, epidemics, direct legislative or administrative interference and so on. However, several of these matters are provided for elsewhere in the SBC, and hence force majeure is included as a general catch-all term to provide for any other event or effect that cannot reasonably be anticipated or controlled and that is not otherwise expressly included in the contract.

Exceptionally adverse weather conditions

In earlier editions of the JCT standard forms of building contract, the term 'exceptionally inclement weather' was used. The present term was introduced in the 1980 editions of the JCT contracts following the unusually long, hot summer of 1976 when many building projects were delayed by exceptionally adverse, but by no means inclement, weather.

Part IV

'Exceptionally' is clearly the important word in this phrase and must be interpreted according to the time of year and the timing of the project anticipated by the contract documents. Thus, if it were known at the time the contract were let that the execution of weather-sensitive work would span the winter period, then if the works were delayed by a 2-week period of snow and frost during January, such weather would most probably not be regarded as exceptionally adverse and would therefore not be a relevant event. However, should the snow and frost persist for a continuous period of, say, 8 weeks, this may well be regarded as being exceptionally adverse and therefore constitute a relevant event giving an entitlement to an extension of time.

It is also important when interpreting the word 'exceptionally' to have regard to the location of the works, for what may be exceptional in the home counties may not be so in the north of England.

SBC procedure in the event of delay

Clearly, it is in the interests of both the employer and the contractor to take all possible steps to avoid delays occurring, but when they do, architects/contract administrators must bear in mind that they may give extensions of time only in respect of delays caused by the relevant events listed in clause 2.29 and they may certify payment of loss and/or expense arising only from the relevant matters listed in clause 4.24. Any other delays or loss and/or expense that may be caused by the employer cannot be dealt with by the architect/contract administrator.

It is worth noting that any delays caused by default of the employer that cannot be dealt with under the contract, for example failure to give possession of the site within the period of deferment stated in the contract particulars, can result in time being 'at large' (see previous sections) and employers losing any right that they may otherwise have to levy liquidated damages.

Best endeavours

SBC clause 2.28.6.1 requires the contractor to constantly use his best endeavours to prevent delay in the progress of the Works and to prevent the completion of the Works being delayed or further delayed beyond the Completion Date as a condition precedent to the architect/contract administrator granting an extension of time. This is a demanding proviso and one that would seem to impose upon the contractor an obligation to do everything possible to achieve the earlier completion date. It has been suggested that this obligation does not contemplate the expenditure of substantial sums of money. However, in using their best endeavours, contractors are expected to employ the same resources as would a prudent person to achieve his own ends, and in doing so, they will incur some cost. It is suggested that if the cause of delay is a relevant matter under clause 4.24, the reasonable cost to contractors of using best endeavours is reimbursable as part of their loss and/or expense – otherwise, it is not a recoverable cost. When the cost is not recoverable, contractors will have to decide the extent of the measures that they are prepared to take in order to reduce their exposure to liquidated damages in respect of any delays they may have caused.

Notification of delay

Whenever it becomes apparent that the progress of the works or any section is being or is likely to be delayed, clause 2.27 requires contractors to give written notice to the architect/contract administrator of the pertinent conditions and the event(s) causing delay. It also requires them, either in that notice or in a further written notice issued as soon after the first as possible, to identify those events that in their opinion are relevant and to give particulars of the expected effects and an estimate of the expected delay in the completion of the works or section in respect of each such event. These particulars and estimates have to be updated whenever necessary or as required by the architect/contract administrator.

If the architect/contract administrator is of the opinion that any of the notified events causing a delay is a relevant event and that the completion of the works or section is likely to be delayed beyond the completion date by reason of that delay, he has to fix a later completion date and notify the contractor accordingly in writing. RIBA Publications has published a pro forma notification of a revised completion date – a completed copy is given as Example 32.1 at the end of this chapter.

New completion dates

The architect/contract administrator should fix a new completion date as soon as it is reasonably practicable to do so and, in any event, within 12 weeks of receiving the required notice from the contractor or, where the period from receipt to the contractual completion date is less than 12 weeks, endeavour to do so by the completion date. In the event that the architect/contract administrator fixes a new completion date, he must state which of the relevant events he has taken into account and the extent, if any, to which he has had regard to variations requiring an omission of work.

When the completion date is adjusted for the first time, it can only be extended to a later date. However, after this has first happened, if the architect/contract administrator issues any instructions that result in the omission of work by way of:

- a variation; and/or
- the omission of a provisional sum for defined work,

then a new earlier completion date may be fixed, provided that, having regard to the work omitted, it is fair and reasonable to do so. However, when fixing an earlier completion date, the architect/contract administrator is not allowed to alter the length of any extension of time required by the contractor and for which a Schedule 2 Quotation has been accepted, nor can a new completion date be fixed that is earlier than the date for completion stated in the contract particulars.

Final adjustment

Not later than 12 weeks after the date of practical completion, the architect/contract administrator is required to finalise the position regarding the completion date by taking one of the following three courses of action (whichever course the architect/contract administrator takes, he must commit his decision to writing and forward it to the contractor):

Fix a completion date later than that previously fixed

If, upon reviewing all previous decisions regarding extensions of time and upon reconsideration of all relevant events (whether specifically notified by the contractor), the architect/contract administrator reaches the conclusion that it is fair and reasonable to fix a completion date later than that previously fixed, he may do so. It will thus be seen that due notification by the contractor of a relevant event is not, in the end, a condition precedent to a retrospective grant of an extension of time, although it is a condition precedent to seeking an extension of time before practical completion.

Fix a completion date earlier than that previously fixed

It is to be noted that fixing an earlier completion date has to be related to the omission of work – it cannot result from the architect/contract administrator reducing previously granted extensions of time, which, with the benefit of hindsight, are thought to be overly generous. As when fixing an earlier completion date prior to practical completion, the architect/contract administrator is not allowed to alter the length of any extension of time required by the contractor in respect of a Schedule 2 Quotation, nor is he allowed to fix a new completion date that is earlier than the date for completion stated in the contract particulars.

Confirm the completion date previously fixed

After the final review, it may be that no adjustment is necessary and the contractual completion date for the purposes of calculating any liquidated damages (see the following sections) and the rectification period can remain either as stated originally in the contract particulars or as adjusted following a relevant event prescribed in clause 2.29.

Duties and decisions

There is little doubt that these procedures impose upon the contractor and the architect/contract administrator not inconsiderable obligations whenever there is a likelihood of delay. When seeking an extension of time, the contractor must be specific and the architect/contract administrator must deal with the matter as and when it arises. The intention of the SBC is quite clear – decisions on extensions of time are to be made as quickly as possible after the occurrence of a relevant event that is likely to cause a delay. There is no place under the SBC for the practice of waiting to see what the delay actually is at the end of the contract and then arguing about whether an extension of time is justified. Such practices, if followed, may well cause the contractor to incur additional costs in accelerating the works to meet an unnecessarily onerous completion date.

Whenever the architect/contract administrator has to review the completion date, it should not be overlooked that the master programme, which the contractor is obliged to provide, may well contain much valuable information to assist in the deliberations.

Perhaps, the greatest difficulty that the architect/contract administrator is likely to encounter in relation to the procedures to be followed in the event of delay is in deciding whether the notice, particulars and estimates that the contractor is required to provide are sufficient to make it practicable for a new completion date to be fixed. In making this decision, the architect/contract administrator should, as always, act fairly and must avoid the use of contrived allegations of insufficiency of notice, particulars or estimate as an excuse for delaying the fixing of a new completion date.

The methods that may be adopted in determining an extension of time are the subject of much debate in the construction industry. Certainly, this is a topic beyond the scope of this chapter to address such are the complexities that it raises. The Society of Construction Law (SCL) has produced a highly regarded document entitled 'Delay and Disruption Protocol', which considers the various delay analysis methods that are available and the circumstances when they should or should not be used. Anyone tasked with deciding an extension of time is urged to read the SCL Protocol.

Reimbursement of loss and/or expense under the SBC

Any disruption or delay caused by default of the employer (or those for whom he is responsible) would ordinarily give the contractor a right to claim damages at common law. In the case of SBC, instead of implied rights, there are express and extensive remedies set out in the contract itself. Since clause 4.23 sets down the procedures to be followed in the event of matters materially affecting the relevant progress (of the works or section) and clause 4.24 provides a fairly comprehensive list of relevant matters to recompense the contractor for losses arising from the employer's acts or defaults, it leaves one to wonder what common law rights might exist beyond those already expressly catered for in the contract. However, this matter is not left to chance, and clause 4.26 states: 'The provisions of clauses 4.23–4.25 are without prejudice to any other rights or remedies which the Contractor may possess'.

If the regular progress of the works is materially affected due to deferment of giving possession of the site (where clause 2.5 is applicable) or by one or more of the relevant matters listed in clause 4.24, then, subject to the timely issue of the relevant notices, the contractor is entitled to be reimbursed any direct loss and/or expense that may thereby be incurred and that is not reimbursable by a payment under any other provision in the contract. These relevant matters are, in fact, the same as the relevant events, recognised by clause 2.29, but resulting from the actions or inaction of employers or their representatives and that are listed earlier in this chapter.

Part IV

If contractors consider that they have incurred or are likely to incur direct loss and/or expense due to deferment or as the result of one of the clause 4.24 relevant matters, they must make written application to the architect/contract administrator as soon as it becomes apparent that the regular progress of the works has been, or is likely to be, materially affected. In such circumstances, contractors should obviously give the architect/contract administrator as much information as they can, but when required to do so, they have to submit to the architect/contract administrator such information as should enable him to form an opinion as to whether direct loss and/or expense has been incurred as a result of the matters cited in the application. Also, upon request, contractors have to submit to the architect/contract administrator or to the quantity surveyor such details as are reasonably necessary to enable them to ascertain the amount of loss and/or expense incurred or being incurred. Any amount so ascertained must be included in the next interim certificate and is not subject to a deduction for retention (see Chapter 30).

It must always be borne in mind that there can be an entitlement to an extension of time without an entitlement to reimbursement of loss and/or expense. Thus, although contractors may be granted extensions of time as a result of a delay caused by one or more of the clause 2.29 relevant events, they are not automatically entitled to reimbursement of direct loss and/or expense unless they can show that the cause of the delay was also a clause 4.24 relevant matter and that it actually (as distinct from theoretically) caused them to suffer direct loss and/or expense that could not be recovered under any other provision in the contract – for example through the valuation of a variation. Conversely, there can be an entitlement to reimbursement of loss and/or expense without an entitlement to an extension of time. This can occur when the relevant event is not likely to cause delay that affects the completion date but, as a clause 4.24 relevant matter, does cause the contractor to suffer loss and/or expense – for example delay to a non-critical element of the works, which causes additional expense due to loss of output.

Liquidated damages

It is a basic principle of English law that if one party to an agreement breaks a term of that agreement causing the other party to suffer financial loss, that second party can claim compensation or 'damages'. However, if it were left to aggrieved employers to claim and prove specific losses on each and every occasion that a contractor failed to meet a completion date, it would be a total waste of time, cost and effort in detailed litigation. It is therefore normal in building contracts to provide for employers to recover 'liquidated' damages, a set sum, from contractors if they fail to complete the works or section by the relevant completion date. Such damages must be 'ascertained', that is they must be a genuine pre-estimate of the likely loss that will be suffered by the employer if there is a delay to the completion date of the works or respective section. The pre-estimate of loss does not have to be totally accurate, but to be acceptable if queried in the courts it should be the result of making what is, in all the circumstances, a realistic attempt at pre-judging an employer's losses likely to arise from any delay between the contractual completion date and the actual date of practical completion. If, in contrast, the employer makes no attempt to pre-estimate his loss and instead puts in a considerable sum (from which he would greatly profit were the contractor to be delayed without excuse), then it is likely that the courts would regard such damages as a 'penalty' and not enforce them.

The SBC makes provision for the recovery by the employer of liquidated damages in clause 2.32. It also requires the insertion of an amount and period for those damages in the contract particulars against clause 2.32. It is important that the entry in the contract particulars is carefully considered. As stated previously, the amount of liquidated damages must be a genuine pre-estimate of loss and applies either to the whole of the works or to a section of the works.

The operation for recovery of liquidated damages under the SBC is summarised as follows:

- If the contractor fails to complete the works or section by the completion date, the architect/contract administrator is required to issue a certificate to that effect.
- Provided that such a certificate has been issued the employer may, if he so wishes, inform the contractor in writing that he may withhold, deduct or require payment of liquidated damages. If the employer chooses to inform the contractor in these terms, this must be done not later than 5 days before the final date for payment of the debt due under the final certificate.
- The employer may then advise the contractor in writing that the employer is to be paid liquidated damages at the rate stated in the contract particulars, or at a lesser rate if the employer so chooses, for the period between the completion date and the date of practical completion. Alternatively, the employer may advise the contractor in writing that such liquidated damages will be deducted from monies due to the contractor.
- If, after the payment or withholding of liquidated damages, the architect/contract administrator fixes a later completion date under the procedures referred to previously, the employer must pay or repay to the contractor any amounts recovered, allowed or paid in respect of the period between the original completion date and the later completion date.

Disputes and dispute resolution

As a consequence of building projects in general being complicated, unique ventures erected largely in the open, on ground the condition of which is never fully predictable and in weather conditions that are even less so, it is not uncommon for disputes to arise during the course of building operations. There are many reasons why disputes occur, but in the main they are caused by the failure of one or more members of the building team:

- to do their work correctly, efficiently and in a timely manner;
- to express themselves clearly; or
- to understand the full implications of instructions given or received.

Part IV

Most disputes are of a minor nature and are settled quickly, fairly and amicably by the building team. From time to time, however, more serious issues come into dispute. When this happens, the building team should make every effort to reach a fair settlement by negotiation – insofar as this might be regarded as adopting approaches outside of the rules of the contract, the employer should be kept informed every step of the way and no settlement should be signed off without the employer's express consent. If this fails, it becomes necessary to use one or more of the dispute resolution mechanisms available – mediation, adjudication, arbitration and litigation. Prior to adjudication under the Housing Grants, Construction and Regeneration Act 1996 (the Construction Act), arbitration was the most common method of resolving a dispute arising from a construction contract. However, since the Construction Act came into force on 1st May 1998, adjudication has become the most frequently used method of resolving a dispute.

Mediation

Mediation is a private, informal process in which one or more neutral parties assist the disputants in their efforts towards settlement. As with negotiation, the resolution lies ultimately with those in dispute – the mediator acts purely as a facilitator and cannot impose his decision on the parties.

Adjudication

A statutory process

Until statutory adjudication was imposed on 1st May 1998, the chief weakness of adjudication in the construction industry was that when push came to shove, it was largely unenforceable. However, in the light of issues of cost and delay that were associated with arbitration and litigation (considered in the following sections), Sir Michael Latham's report *Constructing the Team* (published in 1994) resulted in the Construction Act and Scheme in order to give swift, albeit temporary, justice by making the process of adjudication compulsory in most contracts. This is now the single most common mechanism by which construction disputes are resolved. Most construction contracts now contain adjudication provisions compliant with the Construction Act. If a contract were to contain provisions not compliant with the Construction Act, the Act itself would impose statutory adjudication giving a party to the construction contract the unilateral right to refer any dispute arising under the contract to adjudication.

Adjudication is now a binding process, available to the parties at any time, in which the party referring the dispute will receive a decision within 28 days of the referral irrespective of the complexity of the issues in dispute or the amount of money at stake. The 28-day period can be extended by 14 days if the adjudicator obtains the express approval of the referring party, but it can only be further extended if both parties so agree. The decision is binding and enforceable until the dispute is finally determined by arbitration or legal proceedings or by agreement.

The parties may, however, agree to accept the decision of the adjudicator as finally determining the dispute. Even if the adjudicator makes a mistake in his decision, perhaps as a result of the sheer speed of the process, the courts will readily uphold and enforce it, leaving the disaffected party to raise separate proceedings in arbitration or litigation.

The SBC provisions

For any dispute or difference arising under the contract, SBC Article 7 provides that either party may refer it to adjudication in accordance with clause 9.2. That clause incorporates Part I Scheme for Construction Contracts (England and Wales) Regulations 1998 SI/1998/649 (the Scheme), which is subordinate legislation to the Construction Act (and its partial successor, Part 8 of the Local Democracy Economic Development and Construction Act 2009). Clause 9.2 simply requires that:

- for the purpose of the Scheme, the 'adjudicator' is the person (if any) stated in the contract particulars but, in the absence of a name, the nominating body is identified;
- the adjudicator deciding the dispute or difference must, where practicable, have the appropriate expertise and experience in the specialist area or discipline relevant to the issue in dispute; and
- in instances where tests are required for purported non-compliant aspects of the works, if the adjudicator is not appropriately qualified and experienced, he must appoint an independent expert to advise in a written report as to the subject matter of the tests and dispute.

Notice of intention to seek adjudication

Under the Scheme:

- Any party to the construction contract may give written notice of his intention to refer a dispute (or 'difference' to follow the distinction made in SBC clause 9.2) arising under the contract to adjudication.
- The notice of adjudication must contain:
  - a description as to the parties involved in the dispute and the nature of the dispute;
  - details of where and when the dispute arose;
  - the redress sought from the responding party; and
  - specified contact details for each party for the purpose of giving notices.
- In the case of SBC clause 9.2, the person appointed as adjudicator or the nominating body (the professional body empowered to select the adjudicator) is stated in the contract particulars.
- Any person requested to act as an adjudicator must act in a personal capacity and not be an employee of any of the parties and, in order that the adjudicator is seen to be impartial, must declare any interest, whether it is financial or otherwise, in any matter relating to the dispute or difference.
- When a person has agreed to act as an adjudicator, the referring party must refer the dispute in writing to the adjudicator in 7 days and must send copies of the documents to the responding party.

- The adjudicator may, with consent of all the parties, act on more than one dispute under the same contract at the same time. Also, if all the parties agree, the adjudicator may act on one or more related disputes under different contracts.
- The parties to the dispute(s) may agree to extend the period within which the adjudicator may reach his decision in relation to all or any of disputes, whether connected or not.
- An adjudicator may resign his position at any time should the dispute be the same or substantially the same as a dispute previously referred to adjudication. Should an adjudicator resign his position for this reason, the referring party may start again promptly with a new notice and, if requested and insofar as reasonably practicable, supply the new adjudicator with copies of documents that were made available to the previous adjudicator.
- Should a party to the dispute object to the appointment of a particular person as adjudicator that objection does not invalidate either his appointment or any decision reached under his appointment.
- However, the parties to the dispute may at any time jointly and unanimously revoke the appointment of the adjudicator. Should the parties agree to revoke the appointment of the adjudicator, they will be both jointly and severally liable for the payment of reasonable fees and expenses incurred by the adjudicator before revocation, unless the revocation of his appointment is due to the adjudicator's default or misconduct.
- If the adjudicator's appointment is revoked or he resigns, the issues in dispute must be settled by agreement between the parties or referred under the Scheme to a new adjudicator.

Powers of the adjudicator

Under the Scheme:

- The adjudicator must act impartially in accordance with any relevant terms of the contract and in accordance with the applicable law in relation to the contract, and must do so without incurring unnecessary expense.
- The adjudicator may take the initiative in ascertaining the facts and law necessary for determining the dispute and must decide an appropriate procedure to be followed in the adjudication.
- The adjudicator may:
  - request any party to the contract to supply him with such documents that he may reasonably require to supplement the referral notice or other documents that a party intends to rely upon;
  - decide upon the language to be used in the adjudication and whether translation of any document is required;
  - meet and question any of the parties or their representatives;
  - make any such site visits as he feels are appropriate, subject to any necessary third party consent;
  - undertake any necessary tests or experiments, subject to obtaining relevant consents;
  - obtain and consider such representations or submissions as he requires and, providing he has notified the parties of his intention, appoint experts, assessors or legal advisors to assist him; and

- give directions to be complied with as to the timetable for any deadlines or limits as to the length of written documents or oral representations.
- The parties must comply with any request or direction of the adjudicator in relation to the adjudication.
- Should a party fail to comply with the direction, request or timetable of the adjudicator, without sufficient cause, the adjudicator may:
  - continue the adjudication in absence of the third party or the document/written statement requested;
  - draw such inferences from that failure to comply as circumstances may, in the adjudicator's opinion, be justified; and
  - make a decision on the basis of the information before him, attaching such weight as he thinks fit to any evidence submitted to him outside any period he may have requested or directed.
- The adjudicator must consider any relevant information submitted to him by the parties and make available to the parties any information taken into account when reaching his decision.
- Except to the extent that it is necessary for the purposes of or in connection with the adjudication, the adjudicator and any party to the dispute must not disclose to any other person any information in connection with the adjudication where the party supplying it has indicated that it is to be treated as confidential.
- The adjudicator must reach his decision not later than:
  - 28 days from the notice referring the dispute to adjudication;
  - 42 days from the notice referring the dispute to adjudication if the referring party consents; or
  - any such period exceeding 28 days from the notice referring the dispute to adjudication that both parties have agreed to.
- Should the adjudicator fail, for any reason, to reach his decision within the aforementioned time periods, any party may serve fresh notice of their intention to refer the dispute to adjudication and, if requested by the new adjudicator, provide him with all documents made available to the previous adjudicator.
- As soon as practicable following the adjudicator's decision, he is required to deliver a copy of that decision to each of the parties in dispute.

Adjudicator's decision

- The adjudicator must decide the matters in dispute and may take into account any other matters that the parties agree should be within the scope of the adjudication or which are matters under the contract which he considers are necessarily connected with the dispute.
- The adjudicator may:
  - open up, revise and review any decision made by any person referred to in the contract unless the contract states that any such decision or certificate is final and conclusive;
  - decide that any of the parties to the dispute is liable to make a payment under the contract and decide when such payment is due and the final date for such payment; and

Part IV

- having regard to the terms under the contract relating to the payment of interest, decide upon the circumstances in which, the rates at which, and the period for which, compound or simple interest is to be paid.
- In the absence of any directions as to the time of performance relating to the adjudicator's decision, the parties are required to comply with any decision of the adjudicator immediately upon delivery of his decision.
- The adjudicator is required to provide reasons for his decision if requested by one of the parties.

Effects of the decision

- The adjudicator may require, if he thinks it appropriate, the parties to comply peremptorily with his decision.
- The decision of the adjudicator is binding on the parties until the dispute is finally determined by legal proceedings, by arbitration or by agreement between the parties.
- The decision of the adjudicator is binding (see previous section), and the parties are required to comply with the decision. Should either party not comply, the other party is entitled to commence an action for summary judgment to secure such compliance.
- The parties are jointly and severally liable for the payment of the adjudicator's fee and reasonable expenses. The adjudicator may direct how this is to be apportioned between the parties. If he does not do so, the cost is to be borne equally by the parties.
- Unless given express powers, the adjudicator has no power to award the 'winning' party its costs; as such, each party will bear its own costs.
- The adjudicator is not liable for anything he does or omits to do in the discharge of his functions as adjudicator unless he acts in bad faith.

Arbitration

An alternative to litigation

Arbitration is a process, subject to statutory controls, whereby a private tribunal of the parties' choosing formally determines a dispute. It became a preferred alternative to litigation in the second half of the twentieth century because it allowed the parties privacy and an element of control. Arbitration was also, to begin with at least, a more expeditious option, although the process later became beset with the same problems as with litigation, namely delaying tactics by one of the parties.

Many people in the construction industry still take the view that it is essential to have their disputes heard by someone with a working knowledge of building contracts and the practices that are prevalent in the industry. They often lean towards arbitration as a means of achieving that end – although in doing so they fail to take account of the expertise available within the Technology and Construction Court (TCC), which is almost invariably the forum within the High Court structure that hears building contract matters – see the following section.

Arbitration, therefore, essentially represents a process that is available as an alternative to litigation with the purported advantages of flexibility, economy, expedition, privacy and freedom of choice.

SBC Article 8, where it applies, provides for either party to refer to arbitration any dispute or difference arising under the contract, either during the progress or after the completion or abandonment of the works or after the termination of the contractor's employment, except:

- any disputes or differences arising under or in respect of the Construction Industry Scheme or VAT, to the extent that legislation provides for some other method of resolving the dispute or difference; and
- disputes in connection with the enforcement of any decision of an adjudicator.

This provision is an alternative to Article 9, which, by default, provides for the settlement of disputes or differences by legal proceedings. If the arbitration alternative is chosen, it is to be set down in the contract particulars. When arbitration applies, the relevant provisions are to be found in clauses 9.3–9.8 of the SBC.

When arbitration is stated in the contract to be the chosen method of dispute resolution, it will only be in the most exceptional of circumstances that the courts would consent to hearing a dispute under such a contract. Therefore, the decision to arbitrate must not be taken lightly, as it is likely to close off any subsequent recourse to litigation.

Conduct of arbitration

The provisions of the Arbitration Act 1996 apply to any arbitration under the contract, and any such arbitration is to be conducted in accordance with the JCT 2011 edition of the Construction Industry Model Arbitration Rules (CIMAR) current at the contract base date.

Notice of reference to arbitration

If either party requires a dispute or difference to be referred to arbitration, then that party has to serve on the other party a written notice of arbitration identifying the dispute and requiring him to agree to the appointment of an arbitrator.

- If the parties fail to agree on an arbitrator within 14 days of the service of the notice of arbitration, then either party may apply to the person or body named in the contract particulars for the appointment of an arbitrator.
- If the previously appointed arbitrator ceases to hold office, for any reason, the parties may agree upon an individual to replace him or either party may apply to the appointor for the appointment of a replacement.
- Where two or more related arbitral proceedings fall under separate arbitration agreements, the persons who are to appoint the arbitrator may appoint the same arbitrator for all related proceedings.
- After an arbitrator has been appointed, it may be prudent to have any further disputes between the parties heard by the same arbitrator (possibly decided in the original proceedings). The arbitrator first appointed will already have knowledge of the contract and other such information. Nevertheless, the parties

are not compelled to appoint the same arbitrator as they are free to request or agree an alternative arbitrator to hear any further disputes that arise between them.

Powers of arbitrator

The arbitrator has to determine all matters in dispute that are submitted to him and in doing so he has the power to:

- rectify the contract so that it accurately reflects the true agreement made by the parties;
- require such measurements and valuations as he considers necessary to determine the rights of the parties;
- ascertain and award any sum that ought to have been included in any certificate; and
- open up, review and revise any certificate, opinion, decision, requirement or notice issued under the contract.

Effect of an award

- Subject to an appeal on any question of law arising out of an arbitral award, the award of the arbitrator is final and binding on the parties.
- Either party is at liberty to apply to the courts to determine a question of law arising in the course of a reference.

Litigation

The courts provide the setting for the traditional mode of dispute resolution – namely litigation. However, with the past developments of arbitration and then adjudication, the number of disputes actually determined by the courts had declined compared with those settled by other means. Furthermore, very few of the proceedings that are begun actually result in a full trial and subsequent judgment. It is thought that more than 90% of actions begun in the High Court are disposed of before reaching trial. Whereas reference to litigation had at one point waned it has re-asserted itself as a leading means of dispute resolution.

Most building disputes are heard in the Technology and Construction Court (TCC) whose judges are, almost without exception, appointed from leading counsel and spend a great deal of their time hearing disputes about building contracts. It is therefore worth noting that the TCC judges have built up an expertise over the years, which at least matches anything that arbitration has to offer in addition to which, unlike the set up with arbitration, litigants do not have to pay the judge and his support staff or hire the court room.

In his 1996 report *Access to Justice*, Lord Woolf expressed the opinion that the then current system of litigation was too expensive, too slow, too fragmented, too adversarial, too uncertain and incomprehensible to most litigants. As a result of his review, the Civil Procedure Rules were implemented on 26th April 1999 (replacing the former County Court Rules and Rules of the Supreme Court) and apply to all actions begun from that date. It is worth just a mention here that a key facet of these rules was that they were designed to help ensure that all cases are dealt with justly and in ways that are proportionate to the amount of money involved, the importance of the case, the complexity of the issue and the parties' financial position. Lord Woolf's reforms have been well received and implemented, and there is evidence that parties are now opting for litigation in preference to arbitration.

There are three 'tracks' on which litigation runs:

- small claims track for disputes of less than £10,000;
- fast track for disputes between £10,000 and £25,000 (The fast track is the normal track for claims if the court considers that: (a) the trial is likely to last for no longer than 1 day; and (b) oral expert evidence at trial will be limited to – (i) one expert per party in relation to any expert field; and (ii) expert evidence in two expert fields); and
- multi-track for disputes not allocated to small claims track or fast track.

SBC Article 9 provides that, subject to Article 8, the English courts have jurisdiction over any dispute or difference between the parties that arises out of or in connection with the contract. Legal proceedings, subject always to a party's right under the Construction Act to refer a dispute to adjudication at any point in the contract, is the default position for finally determining disputes or differences under the contract. The contract particulars relating to Article 8 state that if disputes are to be determined by arbitration and not by legal proceedings, it must be positively stated that Article 8 and clauses 9.3–9.8 apply.

Issued by: Ivor Barch Associates
address: Prospects Drive, Fairbridge

Notification of
**Extension
of Time**

SBC

Employer: Cosmeston Preparatory School
address: Fairbridge

Job reference: IBA/94/20

Contractor: L&M Construction Ltd.
address: Ferry Road, Fairbridge

Notification no: 6

Issue date: 4 January 2015

Works: New School Library
situated at: Park Street,
Fairbridge

Contract dated: 14 December 2013

Under the terms of the above-mentioned Contract, this is notification that the
Completion Date for

*Delete as
appropriate

* the Works
*~Section no.~ _____ of the Works

previously fixed as
                14 January   20 15
_____

* is hereby fixed later than that previously fixed,
*~is hereby fixed earlier than that previously fixed,~
*~is hereby confirmed,~

and is now
                28 January   20 15
_____

* The extension of time attributed to each Relevant Event is as follows:

   14 days due to variations to the floor finishes in the hall.

* ~The reduction in time attributed to each Relevant Omission is as follows:~

* ~This decision is made by reason of my/our review.~

To be signed by or for
the issuer named
above
              Signed _R.W.Pipe_____

Distribution	☐ Contractor	☐ Quantity Surveyor	☐ M&E Consultant	☐ CDM Co-ordinator
	☐ Employer	☐ Structural Engineer	☐ Clerk of Works	☐ File

for SBC   CONTRACT ADMINISTRATION FORMS   © RIBA Publishing 2011

Example 32.1   Revision to completion date. Reproduced by permission of Royal Institute of
British Architects.

Chapter 33
An Introduction to Sustainability in Construction

John Connaughton

Sustainable development

Sustainability is now a very important concept for the construction sector and for the built environment. It affects not only the outcome of the construction process – the buildings and infrastructure that the industry produces and maintains – but also the way in which construction is carried out and how construction work is procured. This chapter deals with what sustainability means and how it is typically applied in UK construction. Even within the construction domain, the subject area is already vast and is also changing rapidly. The chapter provides a brief overview of the more important concepts and approaches. Suggestions for further reading are provided.

Key concepts

The basic concept of sustainable development that informs much of government and business thinking is taken from *Our Common Future* published by the United Nations in 1987 (also known the Brundtland Report[1] after one of its principal authors). This defines sustainable development as:

> *Development which meets the needs of the present without compromising the ability of future generations to meet their own needs*

> (Brundtland, 1987)

Brundtland identified three fundamental components of sustainable development that need to be kept in balance: (i) environmental protection; (ii) economic growth (iii) and social equity (the so-called 'triple bottom line'). Although these ideas have

The Aqua Group Guide to Procurement, Tendering and Contract Administration, Second Edition. Edited by Mark Hackett and Gary Statham.

been highly influential in shaping thinking and policy development over the last three decades, they are also ambiguous and open to wide interpretation, and a clear operational definition of sustainability remains somewhat elusive. A fundamental problem is that the environmental, economic and social dimensions of development are each quite different. They are defined, valued and assessed differently by different people, and the idea of keeping them in balance is not at all easy to put into practice.

The importance of the environment and the importance of energy

Partly because of these difficulties, and partly because of growing concerns about the impact of development on the environment, a good deal of thinking and effort has focused on the *environmental* dimension of sustainability. Worries about deteriorating environmental quality have given way to more urgent concerns that human growth and development are causing fundamental changes to the earth's climate and ecosystems, reducing their capacity to support a growing population. While some people still dispute the idea of anthropomorphic ('man-made') climate change, there is a high degree of consensus among the scientific community that concentrations of carbon dioxide (CO_2) in the earth's atmosphere have been caused primarily by human activity and, in turn, are causing average global temperatures to rise One very significant consequence of this 'global warming' is an anticipated rise in sea levels, which threatens large sections of the earth's population who live in low-lying regions. A short summary of the scientific arguments can be found in a recent publication by the Royal Society.[2]

A major source of CO_2 emissions is the production of energy by burning hydrocarbon-based (fossil) fuels – coal, oil and gas. Reducing CO_2 emissions through controlling and managing the energy used in all aspects of life – for industry, for transport, for heating and lighting homes and other buildings – is a key cornerstone of most government policies in the developed world that promotes more sustainable development. At an international level, the Kyoto Protocol of 1997[3] set binding targets for 37 countries and the European Union to reduce their CO_2 emissions (and their emissions of other so-called 'greenhouse gases' that contribute to global warming). Not all industrialised countries – most notably the United States – adopted the Kyoto Protocol, and while there have been subsequent efforts to reach similar agreements at a succession of UN-sponsored international climate change summits around the world, there are currently no binding agreements on the scale of Kyoto. A key point, however, is that energy reduction remains fundamental to reducing global environmental impact.

Sustainability in the built environment

For most industrialised countries, the energy used in buildings accounts for up to half of their total energy use and consequential CO_2 emissions (in the UK, it is some 40%[4]). The construction sector therefore has a hugely influential role to play in designing, constructing and maintaining buildings and infrastructure that are energy efficient and not over-reliant on fossil-based energy sources. Of course, energy use and CO_2 emissions are not the only environmental problems. The construction sector consumes the highest proportion (by mass; some 400 million tonnes per annum) of material resources of any other industrial sector in the UK. It also generates large amounts of waste every year – up to 2 tonnes per person per annum in the UK,[5] much of which is dumped in landfill sites. The production of many construction materials can have harmful environmental consequences – for example, deforestation arising from the unsustainable production of timber; air pollution from the production of steel, cement, glass and plastics. In addition, the construction process itself can give rise to further harmful effects, including water pollution from uncontrolled 'run-off' from construction sites; animal habitat destruction and a loss of biodiversity from site clearance and excavation activities, and so on. More environmentally aware construction processes that minimise material resources, reduce waste and avoid the potentially destructive effects of site activities have a significant role to play in improving the sustainability of construction.

The regulatory framework for construction

There is currently a plethora of policies and regulations at both a national and a European level directed at improving the sustainability both of the construction sector and of the buildings and infrastructure it produces. These are in addition to more general policy, legislation and regulation governing the environmental performance of all industrial and workplace activities, which also cover construction sites and buildings – for example, the Control of Pollution Act 1974 and the Environment Act 1995, which impose constraints on noise, air and water pollution, amongst other impacts. A comprehensive guide is beyond the scope of this Chapter – Atkinson, Yates and Wyatt (2009)[6] provide an overview – but it is important to check with the relevant authority as national and EU policy and legislation is constantly changing. The following sections highlight only the more significant developments directly relevant to construction.

European Union developments

A good deal of relevant EU regulation is now driven by broad policies on climate change and energy reduction. Currently, in its Roadmap 2050[7], the EU has set itself a target of reducing carbon dioxide and other greenhouse gas emissions by 80% (on 1990 levels) by 2050. Interim targets include reductions of 20% by 2020 and 40% by 2030. The UK has implemented similarly ambitious national targets in the Climate Change Act of 2008[8] and has developed a strategy for achieving them in a series of Carbon Plans[9], which are regularly reviewed. While these targets will drive further legislative action across both the EU and the UK, a good deal of the legislation already in place pre-dates Roadmap 2050. Key EU legislation includes the following:

- **The Energy Performance of Buildings Directive**[10] (the EPBD), which is designed to promote more energy-efficient buildings across the EU. First

Part IV

published in 2002, it required all EU states to improve their building regulations and to introduce energy certification schemes for buildings as well as inspection regimes for boilers and air conditioning systems. The EPBD was implemented in the UK by The Energy Performance of Buildings Regulations (2007; recast 2012)[11], and important requirements include the following:

- the provision of *Energy Performance Certificates* (EPCs) to provide an assessment of likely energy performance for non-domestic buildings. EPCs are required when such buildings are constructed, subject to major renovations, sold or rented to a new tenant.
- the provision of *Display Energy Certificates* (DECs) to provide a record of actual energy performance for public non-domestic buildings that is prominently displayed for building users to see; DECs are required for all such buildings over 500 square metres floor area/Proposals to extend DECs to non-public buildings (commercial offices, shops, factories etc.) in the UK have, at time of writing (October 2014), not been taken up by the UK government.[12]
- regular (at no greater than 5-yearly intervals) inspections of air-conditioning systems with an effective rated output of more than 12 kW by an accredited air-conditioning energy assessor.

■ **The Construction Products Regulation**[13] (CPR) has recently been implemented and replaces the Construction Products Directive (CPD). The CPD was originally intended to assist free trade in construction materials and products across the EU and had a strong influence on the harmonisation of product standards. The CPR has stronger provisions on sustainability and, in particular, on the sustainable use of natural resources in the production of construction products. It requires manufacturers to provide environmental product declarations (EPDs) for the environmental performance of their products.[14] Note that the CPR, as an EU Regulation, has legal force throughout every member state once implemented – it does not require separate national legislation.

Under the EU Energy Efficiency Directive (EED), the Government has recently set out proposals for an **Energy Saving Opportunities Scheme** (ESOS[15]), which will require all large organisations in the UK to undertake energy audits, including audits of energy use in buildings. ESOS came into effect in 2014.

UK regulatory and policy developments

Carbon reduction targets in the UK Climate Change Act 2008 set the broad context for UK government regulation and policy in relation to the sustainability of construction and buildings. Key **regulations** that impact directly on construction include the following:

■ **The Planning Act** (2008) and related regulation and guidance, including, for example, the Town and Country Planning (Environmental Impact Assessment) Regulations 1999, which require Environmental Impact Assessments (EIAs) for major developments. Also included in this broad category is the National Planning Policy Framework[16] (NPPF), introduced in 2012 to simplify and replace guidance provided in Planning Policy Statements (PPSs). The NPPF has a strong focus on sustainability and includes a 'presumption in favour of sustainable development', essentially giving priority to proposed development that is demonstrably sustainable and placing a strong emphasis on the need for renewable energy and low carbon buildings.

- **The Building Regulations**[17] (2000), which govern the design and construction of buildings relating to a range of requirements including health and safety, accessibility and, crucially for this chapter, energy efficiency and water efficiency. Part L of the Building Regulations covers energy efficiency. The current version in England at the time of writing is Part L 2013, which came into force in April 2014. Different provisions apply in Wales, Scotland and Northern Ireland. Requirements relating to energy consumption and CO_2 emissions are progressively being strengthened on a 3- to 4-yearly revision cycle. UK Government targets are for all new housing to be 'zero carbon' from 2016 (new non-domestic buildings from 2019), although there remains some debate on the precise definition of zero carbon (see further sections under *Allowable Solutions*).

Under the **Energy Act 2011**, the Government is, at the time of writing, developing details of the proposed *Private Rented Sector Energy Efficiency Regulations* for both domestic and non-domestic building. These regulations will set minimum energy performance standards for letting for property by private landlords and are expected to come into effect in 2018.[18]

In the **policy** arena, an important development has been the publication of **The Low Carbon Construction Action Plan**[19] (2011), a statement of UK government's intention to support a joint Industry/Government action plan to develop a transition to a low carbon construction sector. The **Green Construction Board** (GCB) is a joint initiative between the UK Government and Industry, established to take forward the Low Carbon Construction Action Plan and generally to promote innovative and sustainable growth in the construction and infrastructure sectors. The GCB has produced a Low Carbon Routemap[20] for the Built Environment setting out key actions for these sectors on a timeline to the Government's target of an 80% reduction (on 1990 levels) in carbon emissions by the year 2050. Targets for carbon reduction in the more recent Industrial Strategy for Construction (**Construction 2025**,[21] published in 2013) are in line with the GCB Routemap. The Board has a strong business focus and is keen to promote the idea that a 'green' construction sector can lead to increased business opportunities both in the UK and internationally.

There are also many Government **incentive and penalty schemes** designed to promote energy efficiency and reduce carbon emissions in both domestic and non-domestic buildings. Some of the more significant include the following:

- **The Green Deal**,[22] a scheme for improving the energy efficiency of existing buildings. Up-front improvement costs are repaid over time through electricity bills, and the key principle (called the 'Golden Rule') is that energy savings will pay for increased electricity charges. The scheme is designed primarily for home improvements but also applies to business premises. The Green Deal was

Part IV

launched for the domestic sector in 2013. At the time of writing, levels of take up of the Green Deal in the domestic sector have been lower than expected.

- **The Energy Companies Obligation (ECO)**[23] is a programme funded by the larger energy suppliers and provides additional financial assistance for home improvements for some householders on benefits and/or low incomes and/or living in older properties. It runs alongside the Green Deal and replaced the Carbon Emissions Reduction Target (CERT) and Community Energy Saving Programme (CESP) from January 2013.
- **Feed-in Tariffs (FITs)**[24] are essentially guaranteed payments for electricity generated by small-scale (less than 5 MW) low-carbon installations (typically PV – Photovoltaic – panels) that householders or businesses may install independently of energy suppliers. These payments encourage investment in low carbon electricity generation. FITs work alongside the **Renewable Heat Incentive (RHI)**[25], which is a comparable scheme for the installation of heating systems powered by renewable energy sources, such as biomass boilers, heat pumps and solar thermal installations. The RHI initially applied to non-domestic buildings and was extended to domestic buildings in April 2014.
- **The Carbon Reduction Commitment Energy Efficiency Scheme,**[26] usually referred to as the Carbon Reduction Commitment (CRC), is intended to encourage investment in energy efficiency in existing non-domestic buildings. It requires business and other organisations with a certain minimum electricity consumption to purchase allowances for the CO_2 emissions (primarily from energy use) that they are responsible for, and is essentially a tax on CO_2 emissions. It was initially introduced in 2010 as a complex emissions trading scheme but was withdrawn and re-introduced in 2013 as a more straightforward charge on emissions.

The foregoing provides only a very short summary of some of the significant regulations and policy developments on the sustainability of construction and buildings. The landscape is complex and changes rapidly. Readers are referred to the more detailed references at the end of this chapter for further information and guidance.

Assessing the sustainability of construction and buildings

While there is a good deal of regulation and policy relating to the sustainability of construction and buildings, a wide range of voluntary mechanisms and standards are also highly evident in the day-to-day business of construction. The general area of sustainability has long been seen as fertile ground for innovation in construction, and one such important innovation was the development of a new voluntary environmental assessment method for buildings in the UK in 1990 developed at the Building Research Establishment (BRE). Called BREEAM (BRE Environmental Assessment Method), it was one of the earliest schemes for assessing the sustainability of building projects. BREEAM was intended to promote higher standards by providing an environmental rating scheme using a simple scale that designers, contractors and their clients could adopt voluntarily (further details are provided in the following sections). An important objective of the scheme was to create a desire among users to aspire to higher ratings on the scale as they became familiar with it. BREEAM has been very influential in UK construction and has given rise to similar schemes in other countries and for various different aspects/sectors of construction activity as well. There are now many such schemes around the world, and the following section provide an overview of the more significant ones, concentrating on the UK.

UK building environmental assessment schemes and standards

The UK construction sector is dominated by BREEAM, but there are also other methods. The main schemes include the following:

- **BRE Environmental Assessment Method**[27] (BREEAM), which is essentially a family of building environmental standards and assessment methods covering a range of building types for different stages in the construction/building life cycle. BREEAM assesses the environmental impacts of (mainly new) buildings against a range of criteria, including energy performance, water efficiency, occupant health and wellbeing, waste, land use and site ecology and others. It rates buildings on a five-category scale from 'Pass' (the lowest) to 'Outstanding' (the highest). An 'Unclassified' rating is used to denote non-compliance with BREEAM minimum standards. The main versions of BREEAM include the following:
  - BREEAM for New Construction, covering a range of non-domestic building types.
  - BREEAM In Use, with a focus on existing buildings.
  - BREEAM Refurbishment, with separate schemes for domestic and non-domestic buildings.
  - BREEAM Communities, covering environmental assessment at the neighbourhood (rather than individual building) scale.
  - There are also BREEAM schemes for jurisdictions other than the UK – see under *International* in the following section.
- **CEEQUAL**[28] (Civil Engineering Environmental Quality Assessment and Award Scheme) is an evidence-based assessment method for improving sustainability in civil engineering, infrastructure landscaping and public realm projects. Similarly to BREEAM, it assesses the environmental impacts of projects against a range of criteria that are more specific to civil engineering. It rates projects on a four-point scale from 'Pass' (the lowest) to 'Excellent' (the highest). Versions are provided for UK and Ireland projects, international projects, and it also covers Term contracts.
- **Ska Rating**,[29] an environmental assessment tool for building fit-out work. Promoted by the RICS (Royal Institution of Chartered Surveyors), the Ska Rating assesses fit-out work against a range of environmental criteria and is focused on the *fit out project* at three key stages rather than the completed works. It rates projects on three-point scale of Bronze, Silver and Gold.

Part IV

Since 2006, the **Code for Sustainable Homes** was the national standard for the sustainable design and construction of new housing. It is essentially an environmental assessment method that rates new housing on a six-point scale against a range of sustainability criteria. Following a review of housing standards in 2013, the UK Government appears (at the time of writing) to be moving away from using the Code as a national standard. It may survive – albeit in a changed form – as a voluntary assessment method.

International building environmental assessment schemes and standards

There is a great variety of building environmental assessment schemes in use around the world. While many of them have common features, there are also important differences between them as they tend to be country specific and reflect national characteristics, including climate, environmental priorities, building standards and practices. The most common is LEED (Leadership in Energy and Environmental Design, a U.S. scheme). Some of the more popular schemes include the following:

- **LEED**[30] (Leadership in Energy and Environmental Design) which, similarly to BREEAM, is essentially a family of building environmental assessment methods. LEED was developed in the United States by the U.S. Green Building Council (USGBC), and there are currently multiple versions covering a range of building types as well as 'Neighbourhood Development' (operating at the community/neighbourhood scale). LEED assesses the environmental impacts of new and existing buildings against a range of criteria, including energy performance, water efficiency, indoor environmental quality, materials and resources, site sustainability, and others. It rates buildings on a four-category scale from 'Certified' (the lowest) to 'Platinum' (the highest). While the there are many similarities between LEED and BREEAM, the ways in which the environmental criteria are defined (LEED technical requirements are based on U.S. standards) and scored in each scheme mean that they are not directly comparable. LEED is frequently used outside of the United States – including in the UK – and has formed the basis of many new national schemes.
- **BREEAM International**[31] (see under *UK building environmental assessment schemes and standards* mentioned previously). There are a number of national schemes currently being developed using the BREEAM framework. In addition, BREEAM provides a generic European scheme for commercial buildings and a 'Bespoke' scheme that can be tailored as appropriate for particular countries and building types.
- **Green Star**[32] is an Australian scheme with strong similarities to BREEAM on which it was initially based. It is operated by the Green Building Council of Australia. It can be used for a range of building types and community development and uses a similar set of environmental criteria as BREEAM, although based on Australian standards and practice. Buildings are rated on a three-category scale: 4 Star (Best Practice); 5 Star (Australian Excellence) and 6 Star (World Leading).

Other schemes include Green Globes (in use in Canada and, to a more limited extent, in the United States); Estidama (using a 'Pearl' rating system) in Abu Dhabi; CASBEE (Comprehensive Assessment System for Building Environmental Efficiency) in Japan; and others.

There is also a range of voluntary international standards covering environmental assessment at the company/organisation, project/building and product levels. Important standards include the following:

- **ISO 14001: 2004**[33] (Environmental Management Systems) is a widely recognised international standard for the development and implementation of systems for the management and reduction of an organisation's environmental impacts. The ISO 14001 standard is increasingly being used in construction procurement as means of assessing construction firms' commitments to good environmental management.
- **ISO 21931-1: 2010**[34] (Sustainability in building construction – Framework for methods of assessment of the environmental performance of construction works – Part 1: Buildings) is an international standard providing a framework for assessing the environmental performance of building works, covering new and existing buildings at key stages in the life cycle.
- **European Commission Mandate M350** (Integrated environmental performance of buildings) is an EC mandate for CEN (the European Committee on Standardisation) to develop a suite of standards for the assessment of the environmental performance of buildings. At the time of writing, the Technical Committee (TC350[35]) developing this suite of standards was still working on key elements, including integrating environmental, economic and social assessments. An important development in this work is an emphasis on the need for a life-cycle assessment approach to assessment (also see under *Other Important Issues* in the following section).

Author's comment

The emergence of voluntary environmental assessment schemes and standards for buildings is a relatively recent development. There is little doubt that their increasing use in design, construction and in the procurement of building works has had led to improvements in the environmental performance of buildings and the building process. However, current understanding of sustainability in construction is continually changing, and it is therefore helpful, in particular, to bear in mind a number of important limitations of the voluntary assessment mechanisms currently in use:

- Different national schemes reflect their own national climate, building standards and practices – whilst there are similarities between them, they are not generally comparable with one another.
- While national assessment schemes tend to be developed by sustainability experts using a consensus-building approach, many experts often disagree about what sustainability means in construction, the criteria to be assessed,

Part IV

the assessment standards to be adopted, the weighting to be given to different criteria and so on. There is no single perfect scheme that everyone agrees on.

- Different schemes within the same country can also vary in the criteria to be used and how they are assessed. They can give conflicting results for the same project, and it is important to match the assessment scheme to particular project circumstances and client priorities for environmental improvement.

Sustainable procurement

Key concepts

The concept of sustainable procurement in construction can be viewed in a number of ways. Typically, it is seen as the procurement of more sustainable buildings and infrastructure, that is, sustainability relates mainly to the product or *output* of the construction process. The preceding section (*Assessing the sustainability of construction and buildings*) largely covers the key methods and standards for defining and measuring the sustainability of buildings and infrastructure.

Another way to view sustainable procurement is to look at the *inputs* to the construction process – the human resources (professional, managerial, craft and other people-based resources), materials and equipment used for construction – and whether the production and provision of these inputs is carried out in a sustainable manner. For example, there are standards for the sustainable production of timber (including the Forestry Stewardship Council – FSC[36] – standard) that procurers may wish to stipulate to ensure that timber used in building works comes from managed forests that take account of indigenous peoples' rights, biodiversity and other matters. In addition, and as noted previously (under *International building environmental assessment schemes and standards*), there are environmental management standards, such as ISO 14001, that may be used to help ensure that companies supplying different inputs manage their environmental impacts in a responsible manner.

While the distinction between *outputs* from and *inputs* to the construction process may help procurers define what is important to them and develop more comprehensive procurement approaches that cover all key aspects, such a distinction in practice is not so clear cut. Many of the building environmental assessment methods (e.g. BREEAM and LEED) outlined previously are focused on outputs – they are primarily concerned with the sustainability performance of the completed building in-use – but they also contain input provisions relating to responsible sourcing of materials and other matters. In contrast, schemes such as CEEQUAL and Ska are more input focused, although they also contain provisions for performance in-use.

A third way to view sustainable procurement is in terms of the sustainability of the procurement process itself. This could, for example, cover such issues as to whether the process:

- can be considered fair and equitable to all suppliers;
- seeks to maintain and develop the supplier base; and

- promotes good sustainability practices among supplier companies (including e.g. responsible sourcing of materials, workforce welfare, environmental management).

Guidance and standards

A good deal of the guidance and standards on sustainable construction procurement focuses on inputs to construction (in terms both of the human and material resources provided and of the companies supplying them) and the procurement process itself. There is now a British Standard for Sustainable Procurement[37] (BS 8903:2010) that outlines general principles and procurement procedures that support more sustainable procurement. This standard is expanded upon for use in construction in a detailed guide published by CIRIA (Berry and McCarthy, 2011)[38] that covers the roles of the key players and sets out a seven-step approach for putting BS 8903 into practice.

Important considerations for the sustainable procurement of construction (in terms mainly of the *inputs* to construction) can be viewed in two broad categories – those relating to the supplier's organisation and management and those relating to the supplier's products (including where the constituent materials of these products are sourced):

- **Supplier responsibility and governance,** including the supplier's:
  - Overall strategy/policy in relation to sustainability and Corporate Responsibility (sometimes called Corporate Social Responsibility or CSR).
  - Environmental management systems and processes, such as those included under an ISO 14001-compliant system.
  - Workforce and welfare policies and practices.
  - Procurement policies and systems, including the responsible sourcing of supply chain inputs – see the following sections.
- **Product and supply chain issues,** with a strong focus on the 'responsible sourcing' of raw materials, including the following:
  - The environmental impacts of the supplier's own materials/products, covering a range of impacts in the product life-cycle, including energy consumption, emissions, water use, waste production and other matters.
  - The environmental impacts of the materials/products that provide inputs to the supplier's products.
  - Supplier responsibility and governance matters (as mentioned previously) for other companies providing these inputs (essentially, sustainable procurement requirements should cascade along the supply chain, from one supplier to another).

A BRE Environmental and Sustainability Standard[39] (BES 6001: Issue 2.0) provides specific guidance on the responsible sourcing of building materials and products.

Part IV

Other important issues

This short review attempts to capture the key issues of interest to those entering contracts for the design and construction of sustainable buildings, and for procuring more sustainable construction works in a responsible manner. Inevitably, with such a broad subject area, not all issues are covered; those that are included are not covered comprehensively. The following is a very short summary of some further important issues that the reader may encounter:

- **Life Cycle Assessment** (LCA – sometimes referred to as life cycle analysis) is the assessment of the environmental impacts of producing a product (and 'product' could include a building or infrastructure asset) over its entire life-cycle, from 'cradle to grave'. New standards for the environmental assessment of buildings being developed by TC350 (see previous sections) are adopting a life-cycle approach. Concepts related to LCA include the following:
  - Embodied carbon (or energy), which refers to the CO_2 emissions (of energy consumption) associated with the production/manufacture – rather than the use – of products. The Royal Institution of Chartered Surveyors has recently produced a guide to assessing embodied carbon in construction.[40]
  - Embodied water.
- **Resource Efficiency** – an approach to building design and construction that seeks to optimise the use of physical/natural resources. It is actively promoted by, amongst others, WRAP (the Waste and Resources Action Programme)[41] in the UK. Resource efficiency also seeks to minimise waste at key stages in design and construction, and to promote the re-use and recycling of construction materials and products.
- **Allowable solutions**[42] – a mechanism that allows the expected CO_2 emissions of new buildings that cannot be cost-effectively reduced on site at design stage, to be 'offset' using low or zero carbon measures (such as renewable energy generation) located nearby or more remotely so as to achieve 'zero carbon' new building targets.

References

1. Brundtland, G.H. (ed) (1987) *Our Common Future: The World Commission on Environment and Development*. Oxford University Press, Oxford.
2. Royal Society (2010) *Climate change: a summary of the science*. The Royal Society, London. Available at http://royalsociety.org/policy/publications/2010/climate-change-summary-science/
3. Details of the Kyoto Protocol are available at http://unfccc.int/kyoto_protocol/items/2830.php
4. Department of Energy and Climate Change (2014) *Energy Consumption in the UK*, DECC. London. Available at https://www.gov.uk/government/statistics/energy-consumption-in-the-uk
5. Data available from the Waste and Resources Action Programme are available at http://www.wrap.org.uk/content/resource-efficient-construction
6. Atkinson, C., Yates, A. and Wyatt, M (2009) *Sustainability in the Built Environment: An introduction to its definition and measurement*. BRE Press, Watford.
7. Roadmap 2050: A practical guide to a prosperous, low carbon Europe. are available at http://www.roadmap2050.eu/

8. UK Climate Change Act, 2008 are available at http://www.legislation.gov.uk/ukpga/2008/27/contents

9. Department of Energy and Climate Change (2011) *The Carbon Plan: Delivering our low carbon future*. DECC, London are available at https://www.gov.uk/government/publications/the-carbon-plan-reducing-greenhouse-gas-emissions--2

10. Energy Performance of Buildings Directive are available at http://www.epbd-ca.eu/

11. Recast of the Energy Performance of Buildings Directive are available at https://www.gov.uk/government/publications/improving-the-energy-efficiency-of-our-buildings

12. See, for example are availabel at http://www.ukgbc.org/content/display-energy-certificates

13. Construction Products Regulation, are availabel at https://www.gov.uk/government/publications/the-construction-products-regulations-2013-and-approved-document-7-2013-edition

14. See Construction Products Association (2012) *Guidance note on the Construction Products Regulation, Version 2 – December 2012*. CPA, London are available at http://www.constructionproducts.org.uk/publications/industry-affairs/display/view/construction-products-regulation/

15. Department of Energy and Climate Change (2014) *Energy Savings Opportunity Scheme (ESOS). Guide to ESOS, September 2014* DECC, London are available at https://www.gov.uk/energy-savings-opportunity-scheme-esos

16. Department of Communities and Local Government (2012) *National Planning Policy Framework*, DCLG, London are available at http://planningguidance.planningportal.gov.uk/

17. Details of the Building Regulations, including Approved Documents are available at http://www.planningportal.gov.uk/buildingregulations/

18. Department of Energy and Climate Change (2014) *Private Rented Sector Minimum Energy Efficiency Standard Regulations (Non-Domestic) (England and Wales)* Consultation, July 2014. DECC, London are available at https://www.gov.uk/government/uploads/system/uploads/attachment_data/file/338398/Non-Domestic_PRS_Regulations_Consultation__v1_51__No_Tracks_Final_Version_30_07_14.pdf A link to the Domestic proposals is also available at this address.

19. HM Government (2011) *Low Carbon Construction Action Plan: Response to the Low Carbon Construction Innovation and Growth Team Report*, June 2011 are available at https://www.gov.uk/government/uploads/system/uploads/attachment_data/file/31779/11-976-low-carbon-construction-action-plan.pdf

20. For details of the GCB Low Carbon Routemap, are availabel at http://www.greenconstruction-board.org/index.php/resources/routemap

21. HM Government (2013) *Construction 2025*, July 2013 are available at https://www.gov.uk/government/publications/construction-2025-strategy

22. Details of the Green Deal are available at http://www.decc.gov.uk/en/content/cms/tackling/green_deal/green_deal.aspx

23. Details of the Energy Companies Obligation are available at https://www.gov.uk/energy-company-obligation

24. Details of Feed-in Tariffs are available at http://www.decc.gov.uk/en/content/cms/meeting_energy/Renewable_ener/feedin_tariff/feedin_tariff.aspx

25. Details of the Renewable Heat Incentive are available at http://www.decc.gov.uk/en/content/cms/meeting_energy/Renewable_ener/incentive/incentive.aspx

26. Details of the CRC Energy Efficiency Scheme are available at http://www.decc.gov.uk/en/content/cms/emissions/crc_efficiency/crc_efficiency.aspx

27. Details of BREEAM are available at http://www.breeam.org/about.jsp?id=66

28. Details of CEEQUAL are available at http://www.ceequal.com/

29. Details of the Ska Rating are available at http://www.rics.org/uk/knowledge/more-services/professional-services/ska-rating/about-ska-rating/

30. Details of LEED are available at https://new.usgbc.org/leed/rating-systems

31. Details of BREEAM International are available at http://www.breeam.org/podpage.jsp?id=367

32. Details of Green Star are available at http://www.gbca.org.au/green-star/green-star-overview/

33. Details of ISO 14001 are available at http://www.iso.org/iso/iso14000

34. Details of ISO 21931-1: 2010 are available at http://www.iso.org/iso/catalogue_detail?csnumber=45559

Part IV

35. Details of the work of TC 350 are available at http://portailgroupe.afnor.fr/public_espace-normalisation/CENTC350/index.html
36. Details are available at http://www.fsc-uk.org/
37. British Standards Institution (2010) *BS8903:2010 Principles and framework for procuring sustainably: Guide*. British Standards Institution, London.
38. Berry, C. and McCarthy, S. (2011) *Guide to sustainable procurement in construction*. Construction Industry Research and Information Association (CIRIA), London.
39. Details are available at http://www.greenbooklive.com/filelibrary/BES_6001_Issue_2_Final.pdf
40. The Royal Institution of Chartered Surveyor (2014) Methodology to calculate embodied carbon. *1st edition*. RICS, London.
41. The Waste and Resources Action Programme (WRAP) are available at http://www.wrap.org.uk/
42. Zero Carbon Hub (2013) *Zero Carbon Strategies for tomorrow's new homes*, Zero Carbon Hub, London are available at http://www.zerocarbonhub.org/zero-carbon-policy/allowable-solutions

Chapter 34
Future Trends

Erland Rendall

"You must be the change you wish to see in the world', Mahatma Ghandi observed of the world around him in the late nineteenth century. The construction and property sector have both recently endured a period of prolonged economic pressure, enhanced expectation of efficiency gains as a consequence of technological advancement and culture change. What will the medium to long-term impact of the global financial crisis be for construction? What impact will the polarisation of the market from mega global corporates to the local niche players have on procurement, tendering and contract administration? With the advance of virtual modelling of design and construction, what impact will BIM have across the participants in the design and construction process? After nearly 30 years of frustration around the fragmented, adversarial and inefficient nature of the industry, will effective knowledge management, big data and refined process design really deliver the productivity and quality gains that are expected?

Global -v- local

Over the coming decade, global economics, demographics, security, technology and sociopolitical factors will create a more complex and connected global environment. Low costs will become less of a source of differentiation as globalisation takes hold. Human interaction and relationships will become increasingly valuable and central to competitive advantage through more intense and increased collaboration. Making the journey successfully to these high-value interactions will be a significant challenge. Matching interpersonal skills with productive knowledge will be the key.

The pace of business change over the past decade has been significant and has created new opportunities and challenges alike for any business seeking to operate

The Aqua Group Guide to Procurement, Tendering and Contract Administration, Second Edition.
Edited by Mark Hackett and Gary Statham.
© 2016 John Wiley & Sons, Ltd. Published 2016 by John Wiley & Sons, Ltd.

profitably. The growth of emerging markets in comparison to mature markets that are struggling to maintain their financial performance during the financial crisis evidences a clear two-gear global market. The rise of the emerging economies has given rise to national populations eager for infrastructure, development, new products and services. The Internet has upended traditional industry models and fuelled the accelerating pace of business. Wireless technology has brought office work and conducting business everywhere: the café, the commute, the sofa.

In 2012, 2.5 quintillion bytes of data were created every day according to IBM. Data are now produced and transmitted everywhere – sensors, social media site posts, digital images and movies, purchase transaction records as well as GPS device signals, within this digital age. The advent of big data and corresponding exponential data growth has resulted in 90% of all data in the world having been created between 2010 and 2012. Big data has four dimensions – volume, velocity, variety and veracity. Within the built environment, the variety of data that is present and accessible will generate innovation and fuel change across the sector improving efficiency of decision-making and introducing robust predictive analytics into the planning, procurement and operational phases. Big data is therefore more than simply a matter of size. This reservoir of data, information and knowledge will provide the opportunity to determine new insights through a combination of enhanced hardware and software infrastructure, mobile network access and cloud computing.

The Economist Intelligence Unit's research (*Global Firms in 2020*) predicts that over the next decade, companies of all sizes will expand into new markets and their organisational structures will respond accordingly. The paradox of 'global -v- local' will be maintained with global knowledge, systems and process leverage demanded by clients but with local cultural context and understanding delivered through localised management and resourcing structures.

Shifting demographics will see employees working longer hours, retiring later and experiencing greater personal and family stress due to work. Mobility of workforce will be required to respond to dynamic market conditions. Employment contracts will be constructed accordingly. This next period will therefore see notions of loyalty to organisation significantly reduced. An increasing number of younger people will be engaged in middle to senior business roles but with less experience of previous generations. Despite their numbers reducing, an increased older working population will need to remain in work to afford retirement, as a consequence of living longer and reproducing less.

As cultural diversity increases and technology replaces process workers, corporate culture and soft-skills training will be a focus for training in the coming decade. Knowledge workers will be required to engage and interface with colleagues, peers, customers and clients on a more regular basis establishing trusting relationships whether for frequent short-term project commissions or long-term business relationships. Cross-cultural understanding and tolerance will become an increasingly valuable attribute for both extended team working and complex matrix structures.

This trend leads to the increased challenge of creating and building a truly collaborative culture – both at corporate and at project team levels. Creating meaningful connections within a global organisation of tens of thousands of people spread across myriad countries with significant cultural diversity will be a key challenge to respond to. Technology will play a significant part in the solution, at least for the workforce to work smartly. Corporate cultural definition will also assist in responding to the challenge – visions and their associated pillars will take on new importance as behaviours and traits of those individuals need to be aligned in order to achieve coherence in the global environment.

Global organisations will have the opportunity to tap into the global talent pool bringing new and fresh talent into the organisation through new sources in emerging markets. Engagement with governments, institutions and colleges will provide a rich pipeline of talent in conjunction with Corporate Training facilities being established and enhanced.

Industry and corporate trends

Within any business organisation, automation and process improvement will continue to be a major force for activity. The hunt for competitive advantage will increasingly focus on improving the productivity and performance of knowledge workers. The specialisation and focus trend will continue at both business level and below. Industries will polarise in response to commoditisation. Processes will be outsourced or placed off-shore in accordance with where they are done best and most economically. Products and services will become more personalised. Collaboration inside and outside companies will widen and deepen as internal teams work across time zones and functions, as clients demand ever more of companies, and as companies demand ever more of suppliers. Relationship skills will be at a premium. Technology will help knowledge workers perform better, thanks to new collaboration and communication tools, new ways to harvest, store, analyse and disseminate unstructured data, and decision-support tools that expand and enhance knowledge workers' abilities. Organisations will become flatter and less hierarchical. Employees will enjoy greater decision-making autonomy and will participate more actively in corporate planning.

Organisations will exploit these trends by investing significantly in technology in order to secure the single best way to increase performance of their employees and business – increasing the efficiency of knowledge workers. Knowledge work provides competitive advantage precisely because it involves creativity, innovation and decision-making, all of which are, of their nature, very difficult to automate because of their human as distinct from process aspect. Supporting such knowledge workers as they perform non-routine tasks will require a mix of new 'knowledge management' tools. The need for new ways to collaborate and communicate will be the first component – collaboration areas or e-rooms that gather all documents and content, discussion threads and plans related to projects and so on into a single virtual workspace (e.g. Project Extranets, BIM, etc.).

The second area in which new tools are starting to emerge relates to the handling of unstructured data such as free-form text, which is currently far harder for knowledge workers to access and manipulate than structured data. This capability provides the corporate (internal) Google functionality and data mining capability. With these foundations in place, it becomes possible to deploy the third component and most speculative type of technology: systems that combine information from collaboration spaces, structured databases and unstructured data sources to provide decision-support functions, and even some degree of automation, to

Part IV

knowledge workers. Web 3.0 or semantic web thinking around linked data already indicates the progress being made in this area, whereby complete internal functions can be combined with external-facing sales and operations and the wider external segment, sector or industry.

Opportunities and challenges

BIM

As the early fundamental challenges of BIM are steadily resolved over the next 5 years – Business Case, Return on Investment (ROI), Intellectual Property Rights, Security of Data, Model Ownership, Professional Liability, Software interoperability, Contractual Frameworks, Plan of Work, Market Capability and Capacity – the conversion of industry laggards to full adoption and implementation of BIM will be realised. Progress from Level 1/2 Capability to an integrated approach at Level 3 leading to Integrated Project Delivery (IPD) will be evidenced in the latter stages of this decade. As a topic in its own right, BIM is addressed further in Chapter 21.

The advent of digital modelling together with increased information being embedded within models presents both an opportunity and a challenge to all players within the built environment. Implementing BIM effectively requires significant changes in the way construction businesses work, at almost every level within the value and supply chain. In addition to technology investment, workflows will require revision and ultimately reinvention, whilst employee training and development will need realignment and assessment and reallocation of responsibilities undertaken across both corporate and project team structures.

As businesses execute an effective strategy and methodology for implementing BIM at organisational levels, new opportunities and challenges will be presented. Convergence and interoperability (technological compatibility) of technology and process will provide the foundation on which to build the longer term changes in culture and behaviour required to operate efficiently and effectively at both local and global levels.

As the players and key functions within the market gain confidence and capability around the technology, cultural change will gain momentum as increased interdependent relationships are formed across disciplines and functions across the whole life of the built asset (as a topic in its own right, built asset consultancy is addressed further in Chapter 3). The impact on procurement, tendering and contract administration will be significant. As client requirements determine priorities that inform the procurement strategy and ultimate contract form, so this next period will see a proportionate change in the approach to procuring design, construction and operation of assets as a consequence of increased integration and efficient process enabled by BIM. The current suite of standard contract forms, with JCT forms currently being the most commonly used in the UK, will continue to evolve and expand to incorporate new levels of multi-party contribution to the process with corresponding collaborative structures and approaches required for Level 3 BIM/IPD.

Lean process and procedures

The process of design, fabrication, assembly and use will continue to integrate through a combination of value assessment, legislation, market forces and seeking competitive advantage. With increased access to real-time and robust data, information and knowledge, lean thinking, process and procedures will permeate across all functions and roles in the built environment. Manufacturing industries – automotive, aircraft, technology and so on – have continually revised and rationalised their processes to improve efficiency, effectiveness and improvement through lean thinking. A popular misconception is that the lean process is suited only for manufacturing. This is not true. Lean applies in every business and every process. It is not a tactic or a cost reduction programme, but a way of thinking and acting for an entire organisation.

Many organisations choose not to use the word lean, but to label what they do as their own system, such as the Toyota Production System or the Danaher Business System. Why? To drive home the point that lean is not a programme or short-term cost reduction programme, but, rather, it exemplifies the way in which the company thinks and operates. The word transformation or lean transformation is often used to characterise a company moving from an old way of thinking to lean thinking, and it requires a complete transformation on how a company conducts business. This, in turn, takes a long-term perspective and perseverance.

As the predominantly fragmented, adversarial and inefficient construction industry moves to eliminate waste and inefficiency within the design and construction process, increased integration of lean process and procedures will lead to the leverage, share and exploitation of knowledge within the built environment and enhanced collaboration.

Knowledge management

A structure for managing knowledge is formed around three core areas: Knowledge; Research and Best Practice.

Knowledge is defined as 'the facts or experiences known by a person'. Whatever the form in which it is held, knowledge is a resource that creates value principally for its owner but also for others, once it is mobilised on projects to meet clients' needs. Whilst considering the exponential increase in digital data, the steps to forming knowledge are the following:

Data   = Unorganised facts
Information = Data and Context
Knowledge = Information and Judgement

Therefore, knowledge is information interpreted and applied – it is the product of people sharing, developing and adapting learning for their own purposes. It is an evolving set of know-how kept current by the people who use it and apply it on a regular basis.

Research is broadly described as the establishment of facts and the discovery and qualification of end results. As a process, it may or may not be scientific, but it is seen to be the analysis of data or information to generate effects when results are presented as findings.

Best Practice involves the methods and techniques, refined over time, that describe the optimum way of completing a task or activity that ensures that it is 'world class' and delivers excellence in customer satisfaction.

At its broadest level through an effective knowledge management structure, there is a need to facilitate:

- Connecting people with people – collaborating, assisting, exchanging ideas and experience.
- Connecting people with information – capturing, organising, retrieving and sharing.

Sharing knowledge will enable any organisation or project team to improve business performance by harnessing and applying know-how and by giving people access to the information that they need to do their jobs.

However, you can't manage knowledge – what you can, however, do is to manage and cultivate the environment in which knowledge can be created, discovered, captured, shared, distilled, validated, transferred, adopted, adapted and applied.

With the anticipated migration from process to knowledge working and the embedding of intelligent data and information within Digital Models, the ability to understand the current state as well as the future state of either business-wide, collaborative team or project-based knowledge management is essential. Undertaking periodic 'Knowledge Audits' will provide a marker on which to base action and implementation plans as well as measuring progress.

Knowledge Management 10-Step Checklist

1. Is there a client/customer for this knowledge currently or in the future?
2. Create a specific scope for the asset.
3. Form networks related to the subject.
4. Tag existing material related to the subject.
5. Provide principles/guide.
6. Build a checklist.
7. Emphasise links to people.
8. Validate the guide.
9. Publish – share the tacit and make it explicit.
10. Keep it alive – initiate feedback and ownership (shelf-life).

Behaviours

In the environment of increased and explicit change, the culture and behaviour of all participants engaged in the briefing, funding, designing, manufacturing, assembling and use of built assets will have increased relevance and importance. Inherent within the technological and process advances that BIM will deliver, increased transparency and openness will be created immediately. The consequential impact on individuals within this environment will be fundamental.

Team Workflow Pattern	Description	Illustration	Response
1. NOT a Team Task / Activity	Work and activities are **NOT** performed as a member of the team; they are performed alone outside the context of the team. Work and activities are performed by an individual working **ALONE, NOT** in a team.	Work received by individual  Work leaves individual	1
2. Pooled / Additive Interdependence	Work and activities are performed separately by all team members and work does not flow between members of the team.	Work Enters Team 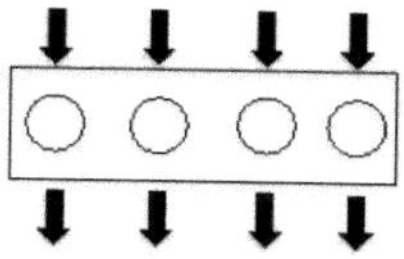 Work Leaves Team	2
3. Sequential Interdependence	Work and activities flow from one member to another in the team, but mostly in one direction.	Work Enters Team 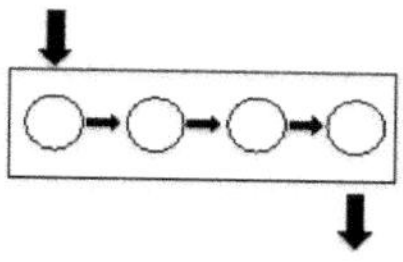 Work Leaves Team	3
4. Reciprocal Interdependence	Work and activities flow between team members in a back-and-forth manner over a period of time.	Work Enters Team 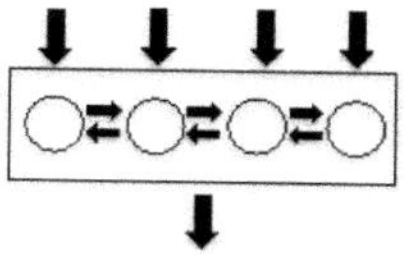 Work Leaves Team	4
5. Intensive Interdependence	Work and activities come into the team and members must diagnose, problem solve and/or collaborate as a team in order to accomplish the team's task.	Work Enters Team 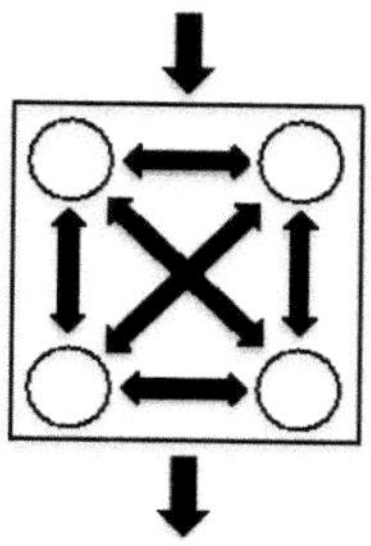 Work Leaves Team	5

Figure 34.1 Team interactivity.

This new virtual and data-intensive environment will demand increased application of knowledge and wisdom, innovation, relationships and cultural awareness. Effective communication skills, across a multitude of channels, will be required as increased global working across cultures, time zones and regulatory frameworks are experienced. Increased interfaces across traditional boundaries will become the norm as the whole-life of the asset is considered and managed within the digital model – from concept to operational facility management and conversion of use.

Enabled by technology and facilitated through revised process workflows and interfaces, a culture of collaboration will increasingly be seen within the construction industry. To be clear, collaboration is a process that can develop when groups or teams work together purposively to solve a problem, make a decision or achieve a common goal, and where members take ownership of and responsibility for the work of the group.

In order to increase collaboration, it will be necessary to:

- improve team cohesion;
- have groups and teams working on tasks where there is an independent goal;
- provide the group or team with regular feedback; and
- give the group or team information about its performance.

Clearly, the task on which a group or team is working, as well as its goal, is important in guiding individual team member behaviour. The task must be one that requires at least some level of interaction and the goal must be a team goal (Figure 34.1).

As you progress from Workflow Pattern 1 through to 5, the level of team inter-activity (or 'Groupness') required to suit the Workflow Pattern increases. It has been found that interdependence improves cohesion and collaboration, and as a result, it increases both member and overall team productivity, ultimately improving the quality of project outcome.

Part IV

Index

Note: page numbers in *italics* refer to figures; numbers in **bold** refer to examples and tables

The Aqua Group Guide to Procurement, Tendering and Contract Administration, Second Edition.
Edited by Mark Hackett and Gary Statham.
© 2016 John Wiley & Sons, Ltd. Published 2016 by John Wiley & Sons, Ltd.

Index